Wireless LAN Systems

For a complete listing of the *Artech House Telecommunications Library*, turn to the back of this book

Wireless LAN Systems

A. Santamaría
F. J. López-Hernández

Editors

Artech House
Boston • London

Library of Congress Cataloging-in-Publication Data

Santamaría, A (Asunción).
Wireless LAN Systems/A. Santamaría and F. J. López-Hernández
Includes bibliographical references and index.
ISBN 0-89006-609-4
1. local area networks (computer networks). 2. Wireless communication systems.
I. López-Hernández, F. J. II. Title
TK5105.7.S263   1993   93-31145
621.39'81–dc20   CIP

© 1994 ARTECH HOUSE, INC.
685 Canton Street
Norwood, MA 02062

International Standard Book Number: 0-89006-609-4
Library of Congress Catalog Card Number: TK5105.7.S263   1993

10 9 8 7 6 5 4 3 2

To Master Craftsmen like Geppetto,
the maker of a mobile, wireless contraption

Contents

Chapter 1
Wireless LANs: An Overview

Francisco J. López-Hernández, A. Santamaría
E.T.S.I. Telecomunicación
Ciudad Universitaria s/n. 28040 Madrid, Spain

1.1 INTRODUCTION

Personal computers have evolved very quickly from the venerable IBM-PC and Apple II, among others, to the high-performance machines of today. This evolution has gone in three directions: processing capacity, size and weight, and price. Portability is one direct result of this process. Not only general-purpose computers, but hand-held terminals designed for specific tasks (store management, hospital data loggers, and so forth) are available at modest prices.

Users have improved their working methods by taking advantage of the new computer capabilities. Portable personal computers are a big success in the information technology market, because users can take their primary information working tools with them.

On the other hand, there has been an increasing development in local area networks (LANs). LANs allow sharing of information in cooperative work, and for desktop computers they offer a fast and reliable connection. Most LANs are based on transmission media consisting of wire, coaxial cable, twisted pair, or optical fibers (wired LANs), and they require costly installations.

The investment for a new LAN installation includes the cost of software, hardware, and cabling. The cabling costs can be as high as 40% of the whole installation, and problems may arise when the network is reconfigured [1]. All the equipment and software can be reused, but the cables are fixed and the cost needed to move them is more or less the same as a new installation. It has been estimated that the total cost to U.S. industry of relocating LAN terminals could reach $5.6 billion by 1990 [2].

This economic projection has caused a growing interest in wireless local area networks (WLANs) that can offer, in principle, portability and lower installation costs. A great deal of research and development have been done (and is being done) to solve the technological problems in changing a reliable cable for an unguided transmission.

Wireless systems can be installed in different environments, such as offices, manufacturing floors, research laboratories, hospitals, or universities, and they offer a set of applications that includes communication between terminals (wireless LANs or mobile robots) and connections to the telephone network (wireless PBX or pocket phones).

Focusing on indoor WLANs, cellular architectures are preferred. The basic idea is to divide the building into cells and establish a wireless link in each cell. The size of each cell depends on several factors: the technology used to implement the system, the environment, the data rate, and so forth.

This chapter presents an overview of WLANs. Section 1.2 includes an analysis and comparison between the two main technologies used to develop these networks: radio frequency (RF) and infrared (IR). Section 1.3 presents a summary of requirements for WLANs, and Section 1.4 presents an analysis of topologies. Several groups are working to find a standard for WLANs. These works are presented in Section 1.5. Section 1.6 includes connection to external networks and, finally, the conclusions are presented in Section 1.7.

1.2 TECHNOLOGIES

There are several methods of establishing a wireless link between two points: ultrasound, carrier currents through main installations, radio-frequency waves, and unguided optical signals. Only two of them (radio and optical signals) are capable of supporting the high-speed data transmission necessary in indoor wireless LANs.

In this section a comparison between RF and IR technologies will be given, with special emphasis on their specific properties and applications. For a detailed analysis of IR systems, refer to Chapters 2, 3, 4, and 5 in this book. For a detailed analysis of RF systems, refer to Chapters 6, 7, and 8.

1.2.1 Radio Frequency

Two different technologies are grouped together: RF and microwave (MW) transmission. We refer to both technologies (RF and MW) when we use the term radio frequency.

In order to implement an RF link, two methods can be used: narrowband and spread-spectrum techniques (see Ch. 6). Narrowband modulation schemes have problems with multipath transmission (Fig. 1.1) and they are very sensitive to interference, so spread-spectrum technology (SST) is preferred. Among SSTs, the most promising is code division multiple access (CDMA). In this case, a single

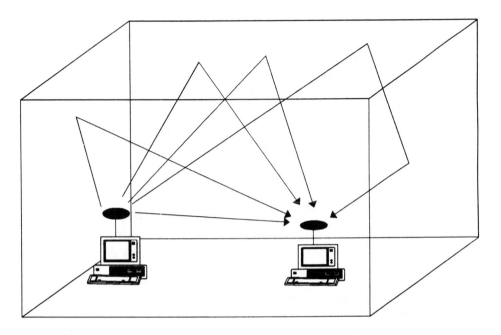

Figure 1.1 Multipath time dispersion.

code is assigned to an emitter inside a cell; if the emitter moves to another cell, a new code is given to continue transmission. The number of simultaneous transmitters has been calculated, and it is larger than that required by other modulation methods [3].

Wireless RF systems present three important problems that must be solved: frequency allocation, interferences and security.

1.2.1.1 Frequency Allocation [4]

There are not too many frequencies free for developing RF WLANs. High-speed data communication is a newcomer to the radio frequency spectrum market, so it has to use spectra that other, older applications are not using. Fortunately, RF WLAN is a short-range application, so the same frequencies can be used by systems that are not too close one another. This policy implies frequency assignment planning for the users, such as user licenses. Standardization groups and users agree that user licenses and frequency coordination should not be required to simplify the development and implementation of WLANs. SST can work with a high degree of interference [5], so if the power emitted is not too high, several neighboring networks could use the same band without problems.

1.2.1.2 Interference

If several WLANs are working in the same building, interference must be avoided. RF signals can penetrate walls; this property is good because a data cell is not restricted to a single room, but it is bad if the neighboring offices have their own networks that perhaps belong to other companies. If no band is assigned to RF WLANs, these systems have to be designed to share the frequency bands, not only with other RF WLANs, but with other services.

1.2.1.3 Security

As RF signals propagate through the walls, data security is an important subject to be considered, and so encryption is mandatory to avoid information leakage. An encryption method has been included in the first European standard for RF WLAN, the Digital European Cordless Telecommunications (DECT) [6] (see Sec. 1.5). If SST is used, greater data security is achieved because only the receiver addressed has the key to decode the data.

On the other hand, RF technology has an older history when compared with other technologies, so it is expected that the research efforts will successfully achieve RF WLANs.

1.2.2 Infrared

Before optical-fiber technology started developing in the mid-sixties, large gas and solid-state lasers were used to transmit information through the atmosphere [7] (see Fig. 1.2). Semiconductor optoelectronic devices such as laser diodes (LD) and light-emitting diodes (LEDs) became useful in the seventies, but then optical fiber was preferable to free air communication. Nowadays, optoelectronic devices working in the 840 to 950-nm range are cheap and reliable. Their main application is in television, wireless headphones, or remote controls.

The most important problems encountered when implementing wireless IR links are the optical power emitted, the multipath intersymbol interference (ISI), and the environmental noise. Focusing on the emitters, we find two kind of devices: *laser diodes* (LDs) and *light-emitting diodes* (LEDs). LDs are fast and powerful devices, but their main disadvantage is spatial coherence. This property is well suited to their use in laser printers and CD players, because it allows small-area

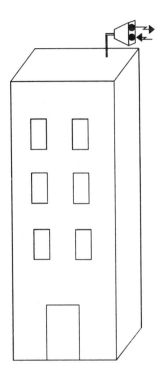 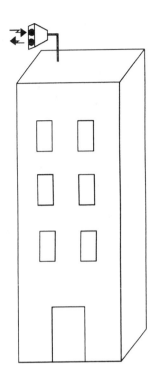

Figure 1.2 Interbuilding optical link.

beam focusing. If light power is confined into a small area, high power densities are obtained. If a laser beam impinges on the eye, severe retina damage may result [8]. So, from the point of view of eye safety, laser emitters are not well suited for wireless indoor applications. For roof-placed interbuilding links, however, it is a suitable device, provided security measures are taken. LEDs are safer than LD for indoor uses. Their main drawback is the power-speed product. That is, there are many high-power (greater than 30 mW) LEDs, and there are many highspeed (less than 10-ns switching time) LEDs, but devices included in both groups are scarce, if not rare. The first group is intended for remote controllers, and the second one is for fiber-optic communication. As the few types common to both groups show, there are few technological problems in making fast and powerful LEDs, but there has been no appealing application for them. WLANs offer just this application. Diffuse links for WLANs will need them, and the potential market for this equipment is great.

Apart from point-to-point links (which will be analyzed in depth in Ch. 5), the main problem that arises on IR WLANs is how to get enough power to the receivers scattered around a room. Two methods have been presented: pure diffuse and quasidiffuse links. In the first method, the optical power is launched in a wide-aperture angle and after several reflections on the room walls it is converted to isotropic radiation, so the receiver orientation is not important since the power comes from every direction (Fig. 1.3). This method was proposed by F. R. Gfeller [9]. Its main drawback is the severe limitation on the channel speed due to multipath propagation. A short emitted pulse will be received, after multiple refections, as a wide one. This effect is known as multipath dispersion and limits the data rate as a function of the room size. Barry et al. [10] have developed a mathematical model to analyze the *room response time*. They conclude that the channel band-width-room length product is roughly 60 Mbps.m, for rooms with 3m-high walls. Then, for a room size of 5m × 5m × 3m, 60 Mbps.m / 5 m = 12 Mbps are expected. If higher rates are used, a severe intersymbol interference is produced by multipath dispersion. The main advantage of pure diffuse links is that the mobility as a line-of-sight between the receiver and the emitter is not needed.

A second method, quasi-diffuse transmission, has been developed to avoid the penalty in speed caused by multipath dispersion [11]. The basic idea is to send

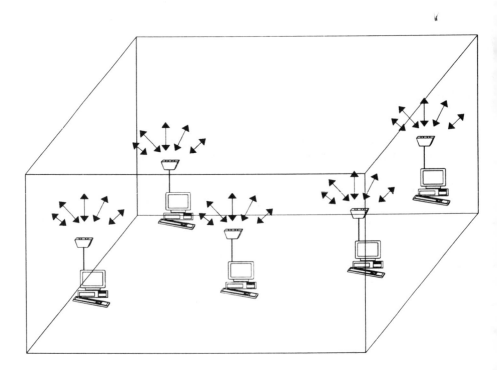

Figure 1.3 Diffuse link.

a thin light beam to a fixed place on the ceiling of the room (satellite), and place all the receivers so they are facing the satellite (Fig. 1.4). If the satellite is a scatterer surface (a small area on a white painted ceiling, for example), we have a passive link. For large rooms, an active satellite is needed (Figure 1.5). In this case, a repeater with several photodiodes and LEDs is used to cover the whole room area. High-power emission is made only by the satellite, so the power needed by the terminals is low. For portable computers, the power requirement is important because they are battery powered and their operational-life time would be impaired by the emitter load. Several active satellites can be used to cover very large rooms. Several point-to-point links are made among the satellites, interconnecting them.

Both active and passive satellites avoid multipath dispersion because the receiver's field of view is small enough to get only the signal coming from the satellite and not from walls. The main advantage of passive satellite is that no installation has to be made for the connection; all the equipment needed is at the terminals. An independent power source for the transmitter would increase battery-life time for portable computers.

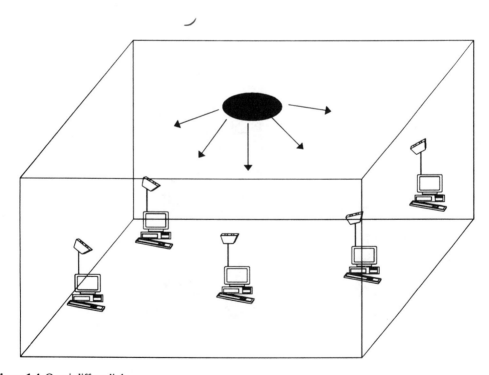

Figure 1.4 Quasi-diffuse link.

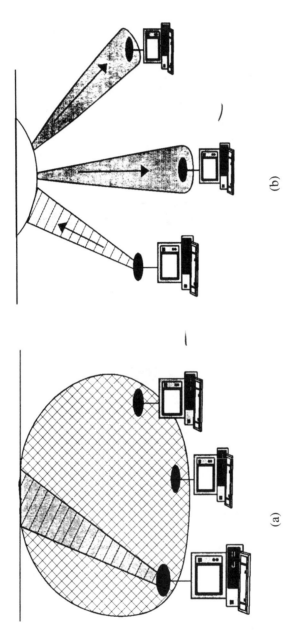

Figure 1.5 Passive (a) and active (b) satellite quasi-diffuse links.

As previously stated, noise and interference due to background light must be reduced in wireless IR systems in order to obtain signal-to-noise ratios that make it possible to establish IR links with low probabilities of error (see Sec. 4.3).

Fast LEDs can be switched at most to around 30-Mpulses per second, so the modulation and coding schemes are restricted to this value. Then, techniques used in RF, such as spread spectrum or CDMA, cannot be used. On-off keying (OOK), multisubcarrier and frequency shift keying (FSK) are the three modulation techniques used in IR wireless links (see Sec. 4.5).

Another problem is the shadowing resulting from moving people. This problem can be avoided by placing both emitter and receiver high enough over the terminal (Fig. 1.6).

There are several advantages of IR wireless links presented as follows. IR signals cannot pass through most objects (walls, doors, etc.), so optical-transmission range is restricted to a room. Then, there is less problem with data security. The only way for information leakage is through windows, but usually the power levels are so low that detection of light crossing windows is almost impossible. In fact, the main problem with optical WLANs is covering the room with enough power, so if a tapping device is placed several meters outside the office, it could not receive enough power to detect the data. Several IR WLANs can be placed in neighboring offices without interference among them. A frequency assignment plan to avoid crosstalk is not needed, so IR wireless transmission is free from Federal Communications Commission (FCC) and other regulations. IR WLANs are well suited for those environments with a high degree of electromagnetic interference. If the

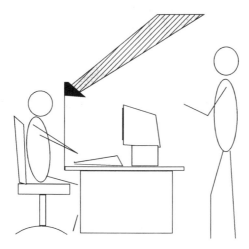

Figure 1.6 Avoiding people-shadowing.

network is placed near a production plant with a high electrical noise, IR technology offers total immunity from electromagnetic interference (EMI) disturbances. Finally, IR equipment is cheaper than RF equipment.

1.3 REQUIREMENTS

In this section, the most important requirements for WLANs are described. These requirements have been classified into: reliability, transparency, throughput, security, mobility, flexibility, topologies, price and, finally, safety and regulations. Several of these requirements have been fulfilled, but others are now under development. Nevertheless, some of the characteristics of WLANs can be relaxed in favor of other characteristics. Although several types of good wireless equipment are on the market, there is a subtle reluctance to their full application. This will happen over the next few years as the reliability of WLANs is proven in real installations.

1.3.1 Reliability

The new medium should be as reliable as a wire system. LAN communication relies on an almost errorless link (probability of error $\leq 10^{-9}$). When an error is detected in a data packet, it is not corrected. The packet is resent. In a coaxial cable or optical-fiber system, the signal-to-noise ratio is large enough to avoid bandwidth-eating self-correcting codes. WLANs must try to keep the error rate at the same level as cabled LANs. This a difficult task to achieve. Wireless systems use lower signal-to-noise ratios (S/N) than cabled links, and when using cellular communication, cell change is an additional source of errors and signal losses. This problem is negligible for voice communication, but it is important for data. DECT, the European standard for wireless communications, states that during cell change the link is kept with both cells on different channels at the same time. There is a close relation between reliability, system throughput, and how the errors are managed. If the error rate is kept low ($< 10^{-6}$), the system performance will be good. Larger error rates should be managed by the hardware, not by the network operating system or drivers.

1.3.2 Transparency

WLANs are not going to replace cabled LANs in indoor environments. They will share the same environment, so the existing software has to work with both types. If the OSI model is assumed, only the first and second layers (physical and link) will be different. Suppose that we want to convert an installed cabled LAN into a

WLAN, keeping the communication cards used to attach each station to the cable, and removing the cable. Interfaces equipment must be used to convert the electrical signals at the communication card to/from wireless (IR or RF) signals. These interfaces must work with total transparency for the user.

For outer environments, where specific standards and products have been developed (at lower data rates than existing LANs), transparency is not required.

1.3.3 Throughput

For the sake of transparency, WLANs should be able to work at the same data rate as cabled LANs. High-speed LANs (FDDI, subDQDB, etc.) are far from the technological possibilities of WLAN. Interactive terminals, where mobility is a goal, can work well in the range 64 Kbps (for text-based terminals) up to 20 Mbps for graphical applications or environments. Then, a hybrid network with medium-speed WLAN cells working with data rates up to 20 Mbps (the user's interface) connected to a high-speed cabled LAN is the best implementation.

1.3.4 Security

Security of cabled LANs is lost when cables are tapped. Using wireless LANs, nobody wants the data flowing around without control. Data encryption is mandatory for WLANs. To avoid degrading the performance, this has to be done by hardware using encryption codes, or by the same method of transmission (using spread-spectrum techniques, for example). For critical applications, terminal (not just user) identification and validation should be made in order to be included in the network. If no security controls are used, the network will be exposed to unfriendly access: jamming packets, airborne viruses, tapping, and so forth. These actions cannot be detected by the lowers levels (physical and link layers in the OSI model), so their control is done by the transport and upper levels.

1.3.5 Mobility

There are two different types of mobility. The first, full mobility, is the ability to send and receive information while moving inside the area covered by the WLAN. This type is difficult to achieve because of the different environments and relative orientations, and because of the distances between network nodes. An example of a full-mobility device is a handheld terminal for security personnel. The second type, weak mobility or portability, is the capability of having a connection to the network by placing a terminal within the area covered by the WLAN, but working at rest. For example, one can place a laptop or notebook computer in a conference room and interchange data and graphs through a WLAN.

Mobility is best understood in a wider communication system, that is, not only data packet transmission but also voice, fax, paging, and so forth. In this context, small terminals capable of unattended (background) communications are needed. As the emitted power is a strong constraint, microcells (600 ft or 200m diameter, or smaller) are an attractive solution for areas with a high density of users.

1.3.6 Network Topology

In WLANs, bus-based topologies are preferred. Nevertheless, if terminals are grouped in clusters and in fixed places, a wireless physical ring can be made using point-to-point links between clusters. These links require precise alignments between transmitters and receivers. Interfaces for Ethernet, token-ring, SNA, and other well-known LANs are offered by WLAN manufacturers, so this is not a very restrictive requirement for the end users. It is important to notice that the whole WLAN is a cell-based network, so previous topologies apply to a single cell.

1.3.7 Flexibility

The number of active nodes in WLANs can change (as in cabled LANs) while the network is working, so the protocols for the inclusion or exclusion of a terminal should be minimized. Several types of networks (mainly token and poll-based networks) need to know how many active nodes are there, so a protocol to get in or out is needed. Others, such as Ethernet, do not mind this and the connection/disconnection process is easy: just place the terminal wherever and start working.

1.3.8 Price

Equipment for WLANs is more complex and more expensive than equipment for cabled LANs, but the advantage of a WLAN is an almost-zero reconfiguration cost. Therefore, for a new wireless network, up to a 100% increase in the price of the wireless equipment in relation to the equipment required to attach a station to a cabled LAN (not including cable costs) would be acceptable now. In the future this difference should be less than 20% compared to cabled LANs.

1.3.9 Safety and Regulations

As intended for office environments, the power levels must be innocuous to human beings and interference with other systems has to be avoided. National and international regulations can make a technology nonapplicable for one defined environment. For example, in many European countries sending radio signals across public areas is not allowed, so the choice for linking two buildings across a city street is to have the connection made by the national PTT or using an infrared beam.

1.4 NETWORK TOPOLOGIES

One WLAN characteristic is that all the nodes share the same medium. Every node hears (although not necessarily listens to) the data flowing around. There are three basic network topologies: ring, bus, and star (see Fig. 1.7).

There are two ways to organize a WLAN: one node per connection, and one cluster of nodes per connection (see Fig. 1.8). In the one-node-per-connection policy, each terminal includes a WLAN interface. This is the most flexible solution, but it is also the most complex. If several terminals are to be placed together (a work team, for example), a cluster distribution is better. In this case, a common WLAN interface is shared by several nodes. Links connecting the interface to the nodes are made with cables. This solution can be a step toward WLAN as a means of keeping the investments made in a cabled LAN.

Any topology can be implemented using any of these configurations, so the following sections are applicable to both strategies. Should any differences exist, they will be noted and explained.

1.4.1 Ring

The ring implies a physical addressing of data from one node to the next. Although an RF or IR signal can be sent in one direction, this is not the idea of a WLAN (in principle, the placement of the nodes is not fixed). Nevertheless, several ring-based IR WLANs have been developed. They are intended for cluster networks and are oriented toward companies that have a token-ring network already installed. To avoid an excessive number of IR point-to-point links, a cluster-based network is used. The basic principle is to control who can send information through the network. There is a special data packet, called a *token*, which gives the right to send data. The information is sent from one node to the next, starting in the origin node, until it is received by the destination node [12].

When a terminal has some information to send, it has to wait until it receives the token from the previous terminal. When this happens, it exchanges the token for its data packet and sends the packet to the next node. So, this cannot do anything but resend the data (it does not have the token), and the process is repeated until the destination gets the message. Again, the same data is sent to the next node. When the circle is closed, the token owner can send another data packet or send the token to the next node.

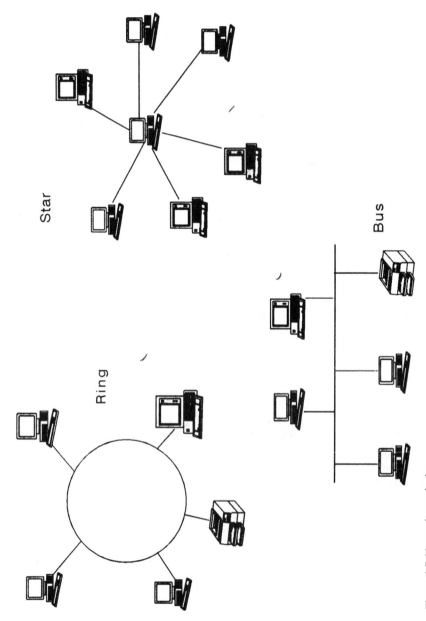

Figure 1.7 Network topologies.

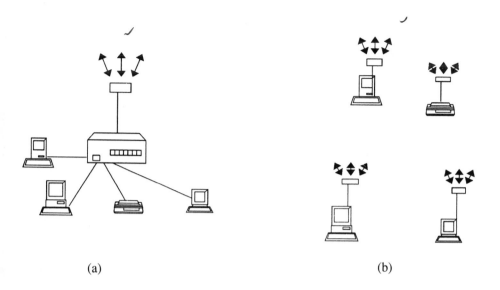

(a)   (b)

Figure 1.8 Cluster (a) and individual (b) WLAN connection.

If a new terminal is connected, a ring reconfiguration is necessary to find out which are the predecessor and successor nodes of the new terminal. If by chance the token is lost, a token recovering protocol is started.

1.4.2 Bus

This is the most *democratic* topology: all the nodes are alike. The problem arises when two or more nodes want to use the channel at the same time. Two methods have been developed to control the medium access: a logical token-passing ring, and carrier detection.

1.4.2.1 Logical Ring

If there is a physical bus network, a logical ring is defined (Fig. 1.9) so every node hears the information, but only *listens* to it when it is its turn. The working mechanism is the same as the previous one (i.e., a token is passed from node to node to enable transmission). Several networks are based on this method. The most well known are: DataPoint's ARCNet, and General Motors' MAP and TOP [13].

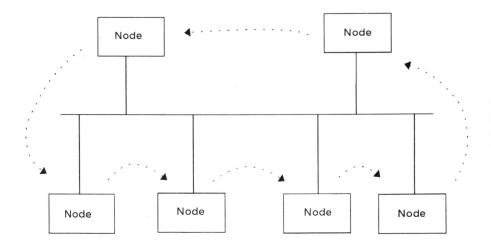

Figure 1.9 Logical ring on a physical bus.

1.4.2.2 Carrier Detection

All the terminals are listening to the channel. Only when no node is talking can a node send a data packet. The problem arises when two nodes start talking at the same time. In this case a collision is produced, and every node notices it. The talking nodes stop transmitting and wait for a random period of time before resending their data. Usually the delay between the two random times is enough to assert that the second node listens to the other, but waits until the first finishes. This is the best method for medium-speed WLANs, that is, up to 20 Mbps. Its main advantage is that the network does not need to know how many active nodes are connected, and so, no reconfiguration protocol is required when this number changes. This method is called carrier sense multiple access with collision detection (CSMA/CD) [14] and its most well known implementation is the ETHERNET, a *de facto* standard for UNIX-based networks.

1.4.3 Star

This is the opposite of bus-based networks. There is a special node (the master) that manages all the information exchange. Other nodes (slaves) can only send and receive data to/from the master. The working protocol most well suited to star networks is called *polling*. The master sequentially addresses the active slaves, and only when a slave is addressed can it send a data packet to the master. If this packet is for another slave, the master will send it when it is addressed. This method is well-suited to networks where the information flux is unbalanced, like in mainframe-to-terminal networks. IBM's HDLC is a good example of this method.

1.5 STANDARDIZATION EFFORTS

In 1985, the Federal Communications Commission (FCC) allocated the so-called industrial, scientific, and medical (ISM) bands for LANs using spread-spectrum techniques [15]. These bands are: 902–928 MHz, 2,400–2,483.5 MHz, and 5725–5850 MHz. No license was needed, subject to no more than 1W being emitted. The covering range is about 800 ft (240m). These bands are not enough for high-speed (greater than 10 Mbps) LANs and new frequencies are under study. These include 17 Ghz and 61-Ghz bands, which are well-suited for RF WLAN [16]. These bands are mainly allocated to radio localization (i.e., radar systems), but the interference between them is difficult: radar systems are insensitive to external signals (in order to avoid deliberate "blinding"), they use high directional antennas and they are placed on rural areas. On the other hand, WLANs are intended to be used in buildings in metropolitan areas. It is important that these bands (17 and 61 GHz) are restricted to these applications and no other service uses them. The FCC has allocated a narrow unlicensed band (1910–1930 MHz) for mobile users, subject to low power emission. Although 20-MHz bandwidth is not enough for very loaded areas like office buildings, hospitals, and so forth, it is a good starting point. If it succeeds, other bands can be allocated for WLANs.

In Europe, the standardization is carried out by the European Telecommunications Standard Institute (ETSI) and the Conference Europenne des Postes et des Telecommunications (CEPT). In March of 1992, ETSI approved the Digital European Cordless Telecommunications (DECT) standard. A new standard for high-performance data networks is under development: HIgh Performance European Radio LAN (HIPERLAN). ISM bands are not available in all the European countries (the 2.4-GHz band is available in Great Britain but not in other countries, for example), because they have been allocated for cellular telephony. Nevertheless, CEPT has proposed the permitting the of use of 2.445 to 2.475 GHz, subject to low power and SST for WLANs. This band is shared with microwave ovens and other short range, high-power equipment. The European Economic Community (EEC) is strongly supporting the development of common standards for all the EEC members, so the differences among national regulations will be reduced in the next few years.

The Institute of Electrical and Electronic Engineers (IEEE) has also focused the presumed necessity of a standard in WLAN. The IEEE standards for LANs have been accepted by International Organization for Standardization and its International Electrotechnical Commission (ISO/IEC), and accepted worldwide. The family of IEEE 802 standards has been a real success in the international standardization of LANs. Within this family, a new working group, known as P802.11, is developing a standard for WLAN. WLAN will share a common medium access control (MAC) and there will be several physical layers (PHY), one for every technology used: ISM-RF, IR, 17-Ghz or 61-GHz bands.

1.6 GOING OUT

Usually a LAN will cover an area larger than just a single WLAN, so the problem is how to interconnect several WLANs to a common-backbone LAN. There is not a unique solution, and the best method is strongly determined by the WLAN topology. For IEEE 802.x networks the interconnection is easy, at least in theory, because of the common-interface network and upper layers. In practice, there are many interface products (bridges and routers) capable of assuring a good translation between networks.

The point is where to make the physical link between the WLAN and the cabled LAN. A node can be used as a bridge to establish the link. In star-based networks, the main node is the choice. For other topologies, any node, in principle, is eligible. The only difference is that this node will manage all external traffic, so it will be addressed more frequently than other nodes. In cluster distributions, the same device (a cluster node connected to external LAN) provides both interfaces: wireless and external communications. For the individual WLAN interfaces, a dedicated (without an attached terminal) node can be used for external communications.

There are two types of external links: intercell or intranetwork links, and external communications. The most important is the first type because it is the way in which a full network is made from independent subnetworks or cells. In this case, full speed and data-frame transparency is needed in order to keep the network performance. The data flow through this interface is large because it is the same network. If the resources are well assigned to the cells, an important reduction in intercell traffic can be achieved, but at least 20% of the data flow will go to other cells. To avoid loop-locks, only one physical path can exist between any pair of nodes in the whole network, or a precise routing method must be established. If the grouping of users in subnetworks is not possible, a hybrid WLAN-cabled LAN will work.

For the second type (i.e., external communications), there is nothing new to say. X25 interfaces, low-speed modems, FDDI connections, or any other gateway devices and protocols can be used from one network node, in the same way as cabled LANs do.

1.7 CONCLUSIONS

There are many cases where the benefits of having a LAN where the nodes are not tied by network cables is worth the cost of a wireless technology [17]. We have seen the most important requirements to be fulfilled by WLANs in order to get a place in the LAN market. Two technologies (RF and IR) can be used to develop these systems under the previously stated conditions, but the efforts and the applicability of both technologies are not equally balanced. RF methods are more complex, but they provide a full mobility and a wide covering range. While IR is easier to implement and no license is needed, its coverage is restricted to a single room or line-of-sight, point-to-point links.

Due to IR limitations and simplicity (both technical and bureaucratic), the largest effort has been dedicated to RF systems. RF WLANs have to share the electrical spectrum not only with themselves, but also with other services using the same frequencies. Several regional or international standards are under development and all of them include a frequency allocation for these systems. Fortunately, the information interchange among these groups is large and all agree to using the same frequencies and modulation schemes, so the chances of getting a common international standard are great [18].

Choosing between IR and RF depends on the network size. For small offices, IR is both the cheapest and the best solution. If a whole building network is to be implemented, RF coverage is better. For industrial applications and machine control, the electrical-noise level makes all the difference between RF and IR. Large industrial plants with low electromagnetic noise at microwave frequencies will be serviced by RF WLANs without any problem, but low-speed diffuse IR communications (using as many satellites as needed) can be the only option for mobile machinery in noisy plants [19].

Table 1.1 shows a comparison between both technologies. Several items, such as price and licensing policy, can change following the market evolution.

In conclusion, WLANs will provide enough options to the LAN designer to suit the user requirements with the optimal solution. Cabled LANs should be kept as the backbone of the network.

Table 1.1
Comparison Between Radio and Infrared Technologies

		Infrared	
Technology	Radio Frequency and Microwave	Diffuse and Quasidiffuse	Point to Point
Mobility	Good	Medium	None
Speed	64 Kbps–1Mbps per channel	1–20 Mbps	Very high
Number of channels	Large	One	One
Licensing	Not for low power and using allocated bands	No	No
Safety	Good*	Good	Possible†
Interference	Possible but avoidable	No	No
Integration of media	Yes	No	Yes‡
Price	Medium to high	Medium to low	Medium to low

*Still under study at larger powers.
†Eye damage with laser emitters.
‡Using higher bandwidth.

REFERENCES

[1] Pahlavan, K. "Wireless Intraoffice Networks," *ACM Trans. Office Inf. Systems*, Vol. 6, No. 3, July 1988, pp. 277–302.

[2] Marcus, M. J. "Regulatory Policy Considerations for Radio Local Area Networks," *Proc. of the IEEE workshop on Local Area Networks*, Worcester, May 9–10, 1991, pp.42–48.

[3] Weber, C. L., G. K. Huth, and B. H. Batson. "Performance Considerations of CDMA Systems," *IEEE Trans. on Veh.*, February 1981, pp. 39.

[4] Baran, N. "Wireless Networking," *BYTE Mag.*, April 1992, pp. 291–294

[5] Pahlavan, K. "Wireless Communications for Office Information Networks," *IEEE Comm. Mag.*, Vol. 23, No. 6, June 1985, pp. 1927.

[6] Davies, W. "From DECT to HIPERLAN the Cordless Revolution in Europe," *Proc. EFOC/LAN'92, Papers on Networks*, Paris , June 24th–26th, 1992. pp. 25–28.

[7] Senior, J. M., ed. *Optical Fiber Communications. Second Edition*. New York: Prentice Hall, International Series in Optoelectronics, Ch. 1, 1992, pp.24.

[8] See Appendix 1.

[9] F. R. Gfeller and U. Bapst. *Wireless InHouse Data Communication via Diffuse Infrared Radiation*. Proc. IEEE, v 67, n 11, Nov. 1979.

[10] John R. Barry, J. M. Kahn, E. A. Lee, and D.G. Messerschmitt. *Simulation of Multipath Impulse Response for Indoor Diffuse Optical Channels*. Proc. of IEEE Workshop on Wireless Local Area Networks. May, 1991. Worcester, Ma. USA. pp. 81–89.

[11] Chu, T. S., and M. J. Gans. "High Speed Infrared Local Wireless Communication, *IEEE Comm. Mag.*, Vol. 25, No. 8., August 1987, pp. 410.

[12] ANSI/IEEE Std. 802.4. *Token-Passing Bus Access Method and Physical Layer Specifications*, The IEEE Inc., February 1985.

[13] ANSI/IEEE Std. 802.5. *Token-Ring Access Method and Physical Layer Specifications*, The IEEE Inc., April 1985.

[14] IEEE 802.3 Std. IEEE/ISO. Reference No. ISO 88023:1989 (E), February 1989.

[15] Davidovici, S. "On the Radio," *BYTE Mag.*, June 1990, pp. 224–228.

[16] Visser, A. R. "European Regulatory Scheme for RLANs," *Proc. EFOC/LAN'92, Papers on Networks*, Paris , June 24th–26th, 1992, pp. 16–21.

[17] Mathias, C. J. "Wireless LANs: The Next Wave," *Data Comm.*, March 1992, pp. 83–87.

[18] Hayes, V. "RadioLAN Standardization Efforts," *Proc. EFOC/LAN'92, Papers on Networks*, Paris, June 24th–26th, 1992, pp. 22–24.

[19] Lessard, A., and M. Gerla. "Wireless Communications in the Automated Factory Environment," *IEEE Network*, Vol. 2, No. 3, May 1988, pp. 64–69.

Chapter 2

Optoelectronic Devices And Circuits

Francisco J. Gabiola, Manuel J. Betancor
E.T.S.I. Telecomunicación
Ciudad Universitaria s/n. 28040 Madrid, Spain

2.1 INTRODUCTION

There are many applications where communication cable wiring is a considerable problem because of environmental considerations. Wireless infrared communications links are an economical alternative for short-distance voice and data communications. Functionally, a wireless infrared system works the same as any other means of communication. A baseband analog or digital signal is modulated and/or codified. After modulation, an electronic circuit drives the light source. This light is then collected, typically using a lens, and transmitted to the wireless medium. On the other side, the light reaching the receiver terminal is collected via a lens, and focused onto the optical detector (photodiode). The photodiode then converts the received optical power into an electrical current, and the next requirement is the transformation of this current into an amplified signal voltage. The signal is fed to the demodulator. Here, the baseband signal information is retrieved from the modulated carrier.

The optical power requirements for light source depends on the type of communication. Focusing on indoor applications, point to point links use light-emitting diodes (LED); however, due to the high optical power levels required in diffuse optical links LED arrays are necessary. We will describe how to build small size arrays using hybrid circuit technology. Light sources suitable for outdoor links are treated in Chapter 5.

On the other hand, the most interesting features of silicon p-i-n photodiodes for indoor applications will be studied.

This chapter includes a discussion of the optical transmitter circuit, with emphasis on circuits for LED arrays. A similar discussion for the optical receiver, including examples of preamplifiers, will also be presented.

2.2 OPTICAL SOURCES

LEDs and lasers are the two semiconductor devices suitable for use in wireless optical communications. A lower optical power, relatively small modulation bandwidth and harmonic distortion are some of the drawbacks of LEDs in comparison to semiconductor lasers. However, light-emitting diodes have a number of advantages that make them more useful for these kinds of communications. The simpler construction of the LED leads to a much reduced cost, which is likely to always be maintained. The light output against current characteristic is less affected by temperature than the corresponding characteristic for the semiconductor laser. Due to the generally lower drive currents and reduced temperature dependence, driver design is simpler.

Since the band gap of the semiconductor material is related to the wavelength of the LED, it is very important to select correctly the material to get the best features in the optical system. Commercially available devices are developed for optical fiber communications; therefore, they are optimized for the three fiber "windows." Two of them are commonly used, one at around 850 nm and the other at about 1,300 nm. Wavelengths in the 1,550-nm range are also exploited in long-haul telecommunications applications, but LED sources in the 1,550-nm range are not commercially available. In wireless links, where the transmission medium is the atmosphere, the window's fiber criterion has to be modified because attenuation and dispersion are conditioned for other elements. To sum up, absorption bands in the near infrared are produced by water vapor and carbon dioxide. However, for small distances (in-house applications) their effects are negligible and a "flat" wavelength spectrum can be considered. Therefore, from the optical-spectrum point of view, devices developed for the three windows of the fiber are useful in wireless in-house optical communications.

The first window's GaAl/GaAlAs LED devices are considered the most interesting option because they can be made with a somewhat simpler process and cost less than second window InGaAsP devices, and because more sensitive detectors are also available for the first window. From fiber terminology, both GaAs and GaAlAs diodes work at the same window; however, due to the different band gap they do not emit at the same wavelength. Gallium arsenide is a direct bandgap material and it has been used for many years as the basis of a number of different types of semiconductor devices. Theoretical wavelength of peak emission is

$$\lambda(\mu m) = \frac{1.24}{E_g(eV)} \approx \frac{1.24}{1.42} = 870 \text{ nm} \tag{2.1}$$

Because of the effect of high doping concentrations on the band edge and the recombination processes, the peak wavelength is slightly longer than this, and may be as high as 930 nm. Gallium aluminum arsenide has a direct band gap over the aluminum composition range $0 < x < 0.37$. From the expression [1]

$$E_g = 1.424 + 1.247 \, x \quad (0 < x < 0.45) \tag{2.2}$$

the corresponding range of values for E_g varies from 1.424 ($x = 0$) to 1.885 ($x = 0.37$), giving a range of values for wavelengths of 650 nm to 870 nm. Infrared commercially available LEDs have a typical wavelength of 850 nm.

Although GaAs diodes have been used for in-house applications, better results are expected with GaAlAs diodes. The use of heterojunctions (e.g., GaAs-GaAlAs) can increase the efficiency resulting from the carrier confinement provided by the layers of a higher bandgap semiconductor (GaAlAs) surrounding the radiative recombination region (GaAs). The heterojunction can also serve as a window for the emitted radiation, because the higher bandgap confining layers do not absorb radiation from the lower bandgap-emitting region. But not only structural features make GaAlAs diodes better than GaAs diodes; as will be shown, the material's intrinsic characteristics are better too.

2.2.1 External Quantum Efficiency

Obtaining a high internal quantum efficiency (η_{int}) is not in itself sufficient to guarantee a successful semiconductor optical source. Photons generated within the LED's *p-n* junction are radiated in all directions, but only a small fraction of them emerge from the surface of the junction to reach the eye of an observer. The ratio of the number of photons finally emitted to the number of carriers crossing the junction is known as the external quantum efficiency (η_{ext}). Four separate loss mechanisms contribute to reduce the quantity of emitted photons (see Fig. 2.1):

1. Only light emitted in the direction of the semiconductor–air interface is useful.
2. Light reaching the emitting surface at an angle greater than θ_c (critical angle) is totally reflected back.
3. Fresnel loss at the semiconductor–air interface.
4. There is absorption between the point of generation and the emitting surface.

The light-emitting region behaves like a Lambertian emitter. In a Lambertian emitter, the power per unit solid angle radiated in directions making an angle θ to the surface normal is $I_0 \cos \theta$.

Therefore, the differential flux can be described as

$$d\varphi = I_0 \cos \theta \, d\omega \tag{2.3}$$

and the differential solid angle has the value

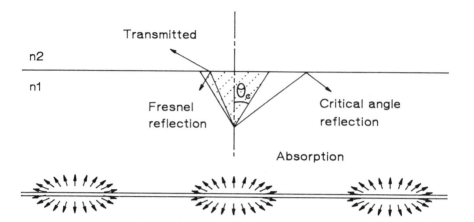

Figure 2.1 Schematic diagram of the main causes of optical losses in a light-emitting diode.

$$d\omega = 2\pi \sin \theta \, d\theta \tag{2.4}$$

The total radian flux from both sides of the generation layer is

$$\eta = \frac{1}{2\pi I_0} \int_0^{\theta_c} I_0 \cos \theta \, 2\pi \sin \theta \, d\theta = \frac{1}{2} \sin^2 \theta_c \tag{2.5}$$

The fraction η of light reaching the surface at an angle smaller than θ_c is

$$\varphi = 2 \int_{\theta=0}^{\pi/2} I_0 \cos \theta \, 2\pi \sin \theta \, d\theta = 2\pi I_0 \tag{2.6}$$

and applying Snell law ($\sin \theta_c = n_2/n_1$)

$$\eta = \frac{n_2^2}{2n_1^2} \tag{2.7}$$

For gallium arsenide LED, whose index of refraction is 3.63, the critical angle is about 16 deg and the fraction of the total power that is able to escape from the semiconductor–air interface is 0.038. For GaAlAs diodes, the index of refraction depends on the aluminum concentration (x) and, in turn, this value conditions the wavelength emission. For x values used in infrared applications, the index of refraction is smaller than the GaAs index, and the critical angle is then greater (about 17 deg). Therefore, efficiency is greater too.

When light passes from a medium whose index of refraction is n_1 to a medium whose index of refraction is n_2, a portion of the light is reflected back at the medium interface. Even those rays within the acceptance angle ($\theta < \theta_c$) suffer some reflection at the semiconductor–air interface because of the change in refractive index. This loss of light is known as Fresnel Reflection. Of the radiation incident perpendicularly onto the surface, a fraction R

$$R = \left(\frac{n_2 - n_1}{n_2 + n_1}\right)^2 \qquad (2.8)$$

is reflected, and only the remaining fraction $T = 1 - R$ is transmitted

$$T = 1 - \left(\frac{n_2 - n_1}{n_2 + n_1}\right)^2 = \frac{4n_2n_1}{(n_2 + n_1)^2} = \frac{4n_1}{(1 + n_1)^2} \qquad (2.9)$$

where $n_2 = 1$ (air).

As is well known, the index of refraction of a semiconductor material can be obtained from the square root of the static dielectric constant ϵ_s. For GaAs diodes $\epsilon_s = 13.18$ and $n_1 = 3.63$; therefore, transmission coefficient is 0.67, whereas for GaAlAs diodes (as we have commented on critical angle) depends on the aluminum concentration through the result of the linear interpolation between GaAs and AlAs [1].

$$\epsilon = 13.18 - 3.12x \qquad (2.10)$$

In general, transmission coefficient is greater since index of refraction is smaller. The overall effect (Fresnel and critical angle losses) according to the semiconductor material index of refraction is presented in Figure 2.2.

2.2.2 Silicon Photodetectors

In the near-infrared region, silicon photodiodes can reach high quantum efficiencies; however, some wavelengths are detected better (efficiency, response speed, etc.) than others. We will now study the sensitivity of silicon photodiodes for GaAs and GaAlAs wavelength. A photodiode has a depleted semiconductor region with a high electric field that serves to separate photogenerated electron-hole pairs. For high-speed operation, the depletion region must be kept thin to reduce the transit time. On the other hand, to increase the quantum efficiency, the depletion layer

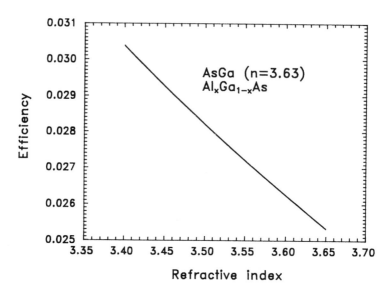

Figure 2.2. Efficiency versus refractive index.

must be sufficiently thick to allow a large fraction of the incident light to be absorbed. Thus there is a trade-off between the speed of response and quantum efficiency.

One of the key factors that determines the quantum efficiency is the absorption coefficient with a strong dependence on the wavelength. For a semiconductor, the wavelength range in which an appreciable photocurrent can be generated is limited. The long wavelength cutoff (λ_c) is established by the energy gap of the semiconductor and has a value of $\lambda_c = 1,100$ nm for Si. For wavelengths longer then λ_c, absorption coefficients are too small to give appreciable absorption. The short wavelength cutoff of the photoresponse comes because of the values of the absorption coefficient for short wavelengths are very large, and the radiation is absorbed very near the surface where the recombination time is short. The photocarriers thus can recombine before they are collected in the p-n junction. Therefore, the range where a semiconductor material is useful as the photodetector is limited by these two wavelengths. The evolution in this range is strongly conditioned by the depletion region thickness (W).

Photodiode performance is more readily optimized by the use of the p-i-n structure, in which the lightly doped n^- layer is made sufficiently thin and is sufficiently lightly doped that the normal bias voltage depletes it entirely. For p-i-n photodiodes, quantum efficiency according to depletion thickness and absorption coefficient is given as [2]

$$\eta = (1 - R)\left(1 - \frac{esp - \alpha W}{1 + \alpha\, l_p}\right) \tag{2.11}$$

where

R = reflection coefficient
α = absorption coefficient
W = depletion region thickness (intrinsic layer)
l_p = diffusion length

For high quantum efficiency, a low reflection coefficient with $\alpha W \gg 1$ is desirable. However, for $W \gg 1/\alpha$, the transit time delay may be considerable, and it may limit the 3-dB frequency. A reasonable compromise between high frequency response and high quantum efficiency for p-i-n photodiodes is obtained for an absorption region of thickness $1/\alpha$. Since the carrier transit time is the time required for carriers to drift through the intrinsic region, the 3-dB frequency is given as

$$f_{3dB} = \frac{2.4}{2\pi t_r} \simeq \frac{0.4\nu_s}{W} \simeq 0.4\alpha\nu_s \tag{2.12}$$

where $\nu_s = 10^7$ cm/seg is the silicon saturation velocity.

Figure 2.3 shows the internal quantum efficiency of the silicon p-i-n photodiodes as a function of the 3-dB frequency and the depletion width. The curves illustrate that at high frequencies the wavelengths corresponding to GaAlAs are detected with better efficiency than GaAs ones.

2.3 LED'S ARRAY

We will now discuss the optical power requirement of the optical source. In point to point channels, distance of 70 meters have been covered at 10 Mbps using commercially available infrared LEDs, such as the L3989 of Hamamatsu [3], and passive optical systems. The distance to be spanned is primarily dependent on the optical aids used. Suitable reflector, lenses, or lens combinations in the beam path of the source and the detector cause a considerable increase in the radiant power falling on the photodetector. In diffused channels, due to the high optical power levels required, an array of LEDs has to be used as optical source. In full-diffused channels, where the goal is to produce a relatively homogeneous diffuse radiation field over the whole room area, the array may be constructed with packed LEDs. However, in quasi-diffused channels, where the optical radiation is collected with a lens or concentrator, the size of the array must be kept small enough to reduce

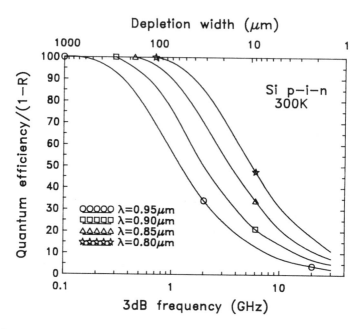

Figure 2.3. Internal quantum efficiency of silicon *p-i-n* photodiodes vs. the 3-dB frequency and the depletion width.

the arraylens coupled losses. Therefore, the array must be constructed with LEDs in chip form. Due to the higher integration density, more power is dissipated in smaller volumes, thus higher peak temperatures. This problem has been solved using hybrid circuit technology with ceramic substrates [4]. The high thermal conductivity of ceramic substrates ($k_{\text{alumina-96}} = 20$ W/°C m^2, $k_{\text{aluminum nitride}} = 170$ W/°C m^2) is a very attractive solution in power electronics, as contrasted with printed circuit boards where expensive cooling fans are needed for the heat removal [5]. The relation between the size of the area for the array and the number of LEDs depends on the thermal characteristics of the single LED , the ceramic substrate, and the environment conditions.

Several mathematical models for thermal simulation of hybrid circuits have been reported in the literature. They are based upon the three mechanisms by which heat may be transferred: conduction, convection, and radiation [6]. The general heat conduction equation can be developed by writing an energy balance for a differential element of a conducting body. However, to obtain useful solutions, certain simplifying assumptions must be made: material homogeneous, isotropic and steady-state conditions. With these assumptions, the differential equation ruling the thermal behavior reads:

$$\left[\frac{\partial^2 T}{\partial x^2} + \frac{\partial^2 T}{\partial y^2}\right] - \frac{T}{L^2} = -\frac{P}{ke} \qquad (2.13)$$

where

k = substrate conduction coefficient
e = substrate thickness
$L = (ke/2h)^{1/2}$: characteristic length
h = convection coefficient (included convection and radiation effects)

This is a Poisson equation with Neumann boundary conditions, so no heat flow out of the borders of the substrate is assumed. As the thickness of commercially available substrates (0.4 mm for aluminum nitride) is much smaller than the other two dimensions, it is reasonable to assume that the temperature gradient between the faces can be neglected, resulting in a two-dimensional model [7].

L can be interpreted as a length constant. Around each LED one can say that (approximately) a zone with radius L will be warmed up. If the distance between two LEDs is smaller than L, their temperature fields will interfere with each other. The numerical solution of (2.13) can be obtained by a finite-difference scheme. Physically, this method is equivalent to the solution of a network with thermal resistances.

Figure 2.4 shows an array of 25 (5 × 5) MFOEC1200 diodes (commercially available in chip form) distributed on a 5 × 5 mm² area, placed on the center of an aluminum nitride substrate. The substrate dimensions are 60 × 60 × 0.5 mm³. The main characteristics of the chip LED MFOEC1200 [8] are indicated in Table 2.1.

If the power dissipation of each LED is 200 mW, the theoretical temperature distribution on the substrate under natural convection can be shown in Figure 2.5. As can be seen, peak temperature is 62°C, and is practically the same for overall array. A comparison between the ideal (not considering thermal effect) and the real (considering thermal effect) optical power obtained from the array is represented in Figure 2.6.

2.4 LED DRIVERS

Having studied light sources for wireless optical communications, we turn to the circuit aspects of operating these devices with emphasis on digital driver designs. A current in the range of 100 to 200 mA must be switched ON and OFF at high speed through the light source in response to a low-level data input signal.

As the applications are too varied, several basic configurations can be pre-

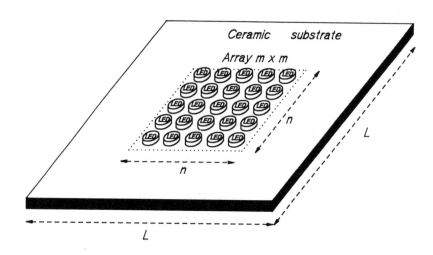

Figure 2.4. Array placement on a ceramic substrate.

Table 2.1
Characteristics of MFOEC1200

Parameter	Value
Forward current peak (1µs pulse, 50% duty cycle)	200 mA
Forward current—continuous	100 mA
Power dissipation	200 mW
Operating junction temperature range	−65 to +125 °C
Forward voltage (I_F = 100 mA)	2V
Total power output (I_F = 100 mA)	2 mW
Wavelength of peak emission (I_F = 100 mA$_{dc}$)	850 nm
Optical rise time (I_F = 100 mA, 10% − 90%)	4 ns
Optical fall time (I_F = 100 mA, 10% − 90%)	5 ns
Die size	24 × 24 mils
Power output versus junction temperature	−0.012 dB/°C

sented. For in-house diffused channels, the multipath causes a spread of the transmitted symbol in time, and the resulting intersymbol interference restricts the digital transmission rate. The theoretical limitation for a room with a length of 10m is 26 Mbps if multipath is the only cause of data rate limitation [9]. Therefore, this application does not require high-speed switching circuits; however, due to the high optical power managed, it must be able to switch ON or OFF an array of

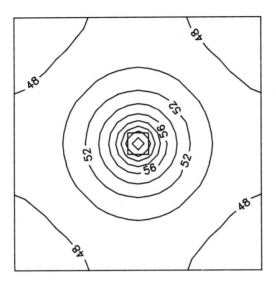

Figure 2.5. Temperature distribution on the array.

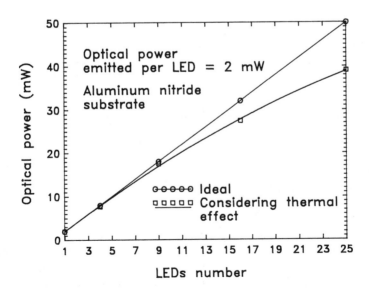

Figure 2.6. Variation of the optical power emitted with the number of LEDs on the array.

LEDs. The two more simple configurations are known as *series driver* and *shunt driver*.

A *series driver* (Fig. 2.7) is a bipolar transistor switch operated in the commonemitter configuration. This circuit provides current gain and a small voltage drop across the switch ($V_{CE\,sat} = 0.3V$). The maximum current flow through the optical source is limited by the resistor R_c. The switching speed of the common emitter configuration is limited by space charge and diffusion capacitance; therefore, bandwidth is traded for current gain. In order to obtain a fast collector response in the active region, the base current must, in any connection, show a strong peak. In the common-base circuit, this peaking occurs automatically if the source impedance is high enough. In the common-emitter connection, this peak must be produced by special means. What appears to be an almost obvious solution to this problem is the generation of a reactively controlled overdrive (preemphasizing) that relaxes under steady-state conditions. In the circuit shown in Figure 2.7 preemphasis is accomplished by use of *speed-up* capacitor C_B [10]. The speed of the common-emitter driver can also be limited by the time required to remove minority-carried charge stored at the collector-base junction during saturation.

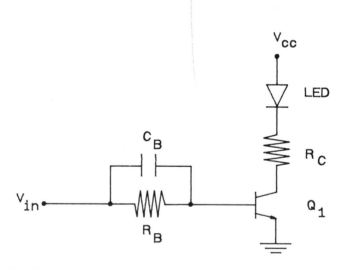

Figure 2.7. Series driver with speed-up capacitor.

This effect can be minimized by using a Schottky diode clamp between the collector and base of the drive transistor. This clamping is automatically provided by use of certain commercially available TTL (transistor-to-transistor logic) integrated circuits. Higher current levels can be handled using Darlington (Figure 2.8) instead of a single transistor.

The user of a Darlington has two choices: either operating the transistor at a current level approaching their defined $I_{C\,sat}$, thus benefiting from the multiplying effect of the gains and having a very high gain switch; or operating the transistors at a higher current level. The forced gain of the Darlington will be very much reduced, but will result in devices capable of switching at much higher current levels than their defined current values. In the simplified Darlington circuit, the emitter current of transistor Q_1 is completely injected into the base of Q_2. Consequently, the leakage current of transistor Q_1, when cut off, is amplified by the transistor Q_2; the result being an overall higher leakage current for the pair of transistors. This effect can be reduced by means of stabilizing resistors between base and emitter of the transistor (Figure 2.8). The influence of R_2 is predominant. The value of these resistors is equally important when considering the storage time. The value of R_1 influences the storage time t_{s1} of the first transistor Q_1. If its value is reduced (all other parameters remaining constant) the storage time t_{s1} is reduced, and the total storage time ($t_s = t_{s1} + t_{s2}$) too, because t_{s2} does not vary. Resistor R_2 influences the storage time t_{s2} of transistor Q_2. The use of a very low resistance value enables the total storage time of the Darlington to be considerably reduced. The use of low-value resistors for R_1 and R_2 is a simple and effective method of reducing the storage time. It should be noted that the use of a low value of R_2 means that the transistor Q_1 has to provide more current. However, the power dissipated in this resistor remains low since the voltage across it is V_{BE2}. In the switching mode, the fall time of the collector current is of great importance, since the majority of the turn-off switching losses take place during this time. For a

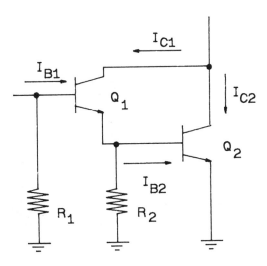

Figure 2.8. Darlington with stabilizing resistor.

Darlington, the fall time of the collector current is only dependent upon transistor Q_2 since transistor Q_1 is already cut off. This dependence is essentially on the waveform of the reverse base current of transistor Q_2.

The *shunt configuration* is one of the two types of low-impedance drivers. The implementation of this circuit with a standard TTL integrated circuit is shown in Figure 2.9.

The switching transistor is placed parallel to the optical source, providing a low-impedance path for switching OFF the optical source by shunting current around it. With DATA IN low, the output transistor is ON, and the current from R_1 is taken to ground through D_1, making the voltage at the anode of the optical source low enough to hold it OFF, yet allow it to be slightly forward biased (by the forwarded voltage on D_1) so it can be turned ON with little delay. With insignificant forward current in the optical source, C_2 is discharged. When DATA IN is raised, the output transistor is turned OFF, allowing the current from R_1 to enter the optical source; there is an initial rush of current as C_2 charges, thus peaking the optical source turn ON. In steady-state ON, optical source current is limited by the sum of R_1 and R_2, but during turn ON current is limited by R_1 only, so the peak-dc current ratio is approximately $(R_1 + R_2)/R_1$. During turn OFF, until C_2 is partly discharged, the voltage on C_2 will apply a small reverse voltage to the optical source, thus peaking its turn OFF as long as the voltage on C_2 remains higher than the voltage at the anode of D_1.

The shunt driver offers other advantages. By replacing the resistor R_1 with a constant current source, power supply noise can be reduced. Furthermore, tem-

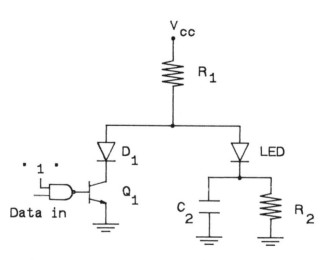

Figure 2.9. Shunt configuration.

perature compensation can also be added by using a temperature-variable constant current source such as that shown in Figure 2.10 [11].

The temperature dependence of the *p-n* junctions of transistor Q_1, Q_2 and diodes D_1, D_2, . . . , D_n is used to increase current with rising temperature

$$I_{\text{LED}} = \frac{R_1}{R_3(R_1 + R_2)} [V_{cc} - 3V_{BE_{\text{diode}}}] - \frac{2V_{BE_{trt}}}{R_3}$$

$$\frac{\partial I_{\text{LED}}}{\partial T} = -\frac{3R_1}{R_3(R_1 + R_2)} \left[\frac{\partial V_{BE_{\text{diode}}}}{\partial T}\right] - \frac{2}{R_3} \frac{\partial V_{BE_{trt}}}{\partial T}$$

(2.14)

where the temperature variation of the cut-in voltage of a silicon junction is -2.5 mV/°C. In series drivers, this is done by its own configuration.

The optical source may be a single LED or an array of LEDs. Several practical circuits to drive a single LED can be found in the bibliography. For an array of LEDs, a 1-Mbps series driver has been implemented that switches 12 diodes in series [12]. Figure 2.11 shows a shunt driver that switches 12 LEDs in series at rates from dc to 12.5 Mbps [13].

In point to point applications, where the medium does not restrict the digital transmission rate, higher speed can be achieved. Increased switching speed may

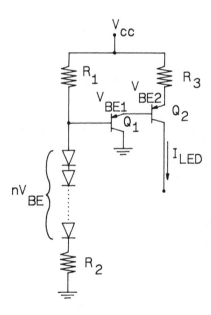

Figure 2.10 Temperature-variable constant current source.

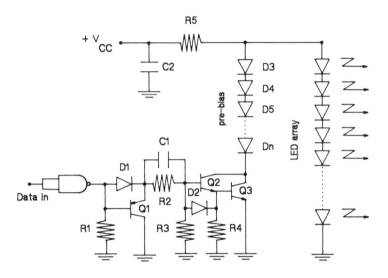

Figure 2.11 12.5-Mbps shunt driver for an array of LEDs.

be obtained with the *emitter-follower drive circuit* and with the *emitter-coupled circuit.*

Emitter follower is the other type of low-impedance driver. This circuit is capable of giving optical rise times of 2.5 ns for LEDs with capacitance of 180 pF, thus allowing 100Mbps operation. The emitter follower is inherently a feedback device. The feedback is normally degenerative at low frequencies. Under certain conditions (usually the existence of drive inductance and load capacitance) the feedback can become regenerative at high frequencies, and cause damped or sustained oscillations. If oscillations are a problem, the use of a parallel RC (resistor-capacitor) network in series with the drive source greatly enhances the transient response. The choice of the resistor and the capacitor depends upon the transistor and the external circuit parameters [14].

An alternative important drive circuit configuration is the emitter coupled or current routing circuit. Although this circuit resembles a linear differential amplifier, it is operated outside the narrow range of linearity in the switching mode of operation. The usefulness of the emitter coupled circuit as a high speed principle is based on its nonlinear application, whereby one of the two transistors is cut off in the steady state condition. The transistor can be strongly overdriven while at same time remaining out of saturation. The effective overdrive results in switching speeds faster than a common emitter circuit would exhibit were it driven from a comparable step input signal. Emitter coupled drivers have been used at speed beyond 300 Mbps [10].

2.5 PHOTODIODES

As we have seen, the choice of the photodetector material is conditioned by the wavelength at which the system works, and for wireless optical communications better results are expected to be obtained with silicon *p-i-n* photodiodes. These devices are characterized by the following parameters:

- Responsivity (A/W),
- Dark current (nA),
- Photosensitive surface (mm²),
- Spectral noise current density (pA/$\sqrt{\text{Hz}}$),
- Capacitance (pF), and
- Response time (ns).

All are largely documented, and therefore we will not include them in this chapter. However, photosensitive surface and capacitance need some comment because of their special incidence in diffused channels. The optical power incident on a photosensitive area A_R for an idealized diffused-optical channel is given as [9]

$$P_R = wA_R \sin^2(\text{FOV}) \quad 0 < \text{FOV} \leq 90° \tag{2.15}$$

where w is radiant emittance and FOV is field of view.

As can be seen, the optical power detected increases proportionally with the photodiode area A_R [15], but capacitance increases to:

$$C_d = \frac{\epsilon_0 \epsilon_r}{w} A \tag{2.16}$$

where A is the junction area, ϵ_0 the free space permittivity, ϵ_r the relative permittivity and W the lightly doped n^- layer. The main characteristics of two commercially available photodiodes are compared in Table 2.2 [16].

C30810 is a large-area photodetector and C30831 is a small-area photodetector. The difference between both capacitances can be observed. As will be shown

Table 2.2
RCA Silicon *p-i-n* Photodiodes

Type Number	Photosensitive Surface Diameter (mm)	Dark Current (nA)	Capacitance (pF)	Response Time (ns)
C30810	11.4	300	70	12
C30831	0.5	10	2	3

later, in order to improve the frequency response it is desirable to choose a photodiode with low capacitance.

A_R in (2.15) can be increased by use of passive optical systems (lenses or concentrators) where the optical radiation is collected on a small-area, low-capacitance photodiode. Practical concentrating systems resemble the theoretical performance for a relatively small angle FOV. Therefore, in full-diffusion channel this is a constraint on the freedom of angular position of the photodiode with respect to the environment. However, a small angle FOV is very useful in Q-diffusion, where only a portion of the ceiling is the Lambertian diffuser. Furthermore, reception noise will be reduced, because the most important noise sources (tungsten lamp, fluorescent lamp, etc.) are placed on the ceiling.

2.6 OPTICAL PREAMPLIFIER CIRCUITS

After the photodiode has converted the received optical power into an electrical current, the next requirement is the transformation of this current into an amplified voltage signal. Together, these two elements (photodetector and preamplifier) dictate many of the receiver characteristics (sensitivity, dynamic range, etc.) and its performance. The rest of the circuitry is more or less standard and has been implemented in many operational communication systems.

As the applications of optical communication systems become more diversified, the optical receivers will have to meet many different requirements. Depending on their configuration, preamplifiers for optical receivers can be classified into two types: high impedance (Fig. 2.12(a)) and transimpedance (Fig. 2.12(b)) [17].

The frequency response of the high impedance preamplifier before equalization is given as

$$\frac{V_A(f)}{I_d(f)} = \frac{AR_f}{1 + j2\pi f R_f C} \tag{2.17}$$

where C is the total input capacitance, including the detector and preamplifier capacitance. R_f is chosen to be large in order to reduce its noise contribution. Consequently, the input time constant $R_f C$ is very large compared to the bit interval T, and the amplifier tends to integrate the signal. An equalizer is necessary to extend the receiver bandwidth out to the desired value. If the equalizer is a passive RC network, the preamplifier gain A has to be large enough to ensure that noise sources from amplifying stages after the equalizer do not degrade the signal to noise ratio. Thus, the maximum voltage swing at the preamplifier output, particularly for low frequency components, is limited by the high value of R_f and A, and the dynamic range of the receiver preamplifier, defined as the ratio of maximum and minimum allowable input signals, is limited too.

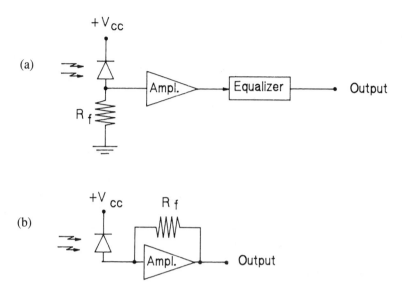

Figure 2.12 Basic optical preamplifier design: (a) high impedance, (b) transimpedance.

Due to the small angle FOV used in Q-diffusion channels, the received optical power is dependent on the position and angular orientation of the photodetector with respect to the Lambertian diffuser surface. Therefore, in these kind of channels there is a wide range of received optical power levels, and a wide dynamic range design (as the transimpedance configuration) is required.

The 3-dB bandwidth of the transimpedance preamplifier is given as

$$f_{3dB} = \frac{A}{2\pi R_f C} \tag{2.18}$$

where C includes the photodetector and preamplifier capacitance, and feedback stray capacitance has been neglected. In this expression we can see how important a low photodetector capacitance value is. For the transimpedance design, no equalization is required because the input-time constant is only $R_f C/A$, a factor A smaller than the inputtime constant of the high-impedance design. Therefore, this configuration offers a higher dynamic range than the high-impedance design. The most important parameter of the transimpedance configuration is the feedback resistance value, since it affects the amplifier bandwidth, gain, dynamic range, and noise (sensitivity).

A great number of optical receiver preamplifiers using BJT and/or FET devices have been implemented. The applications are too varied, and the best choice for 200 Mbps, for example, might not be the best choice for 10 Mbps. For wireless optical networks, a basic feedback configuration is used because the maximum data rate (12.5 Mbps for optical Ethernet) is very low. Figure 2.13 shows a convenient circuit for these applications.

These feedback designs are normally described or classified by the type of feedback associated with their input and output ports. Therefore, Figure 2.13 is a shunt-shunt feedback circuit. Because of the shunt feedback associated with the input and output stages, the circuit offers a low input and output impedance, and makes this amplifier a good transimpedance amplifier. The gain and phase versus frequency of this circuit for a feedback resistance R_F of 12 kΩ are shown in Figure 2.14. If the bit rate of the application is lower, R_F can be increased and a higher value of transimpedance will be obtained.

2.7 CONCLUSIONS

In this chapter we have studied the optoelectronic devices and circuits used in wireless optical communications. Better results are expected to be obtained with GaAlAs diodes and silicon *p-i-n* photodiodes. However, with commercially available LEDs the optical-power levels required in diffused channels cannot be satisfied, and LED array has to be used.

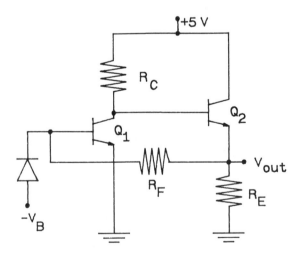

Figure 2.13. Transimpedance design with bipolar transistors.

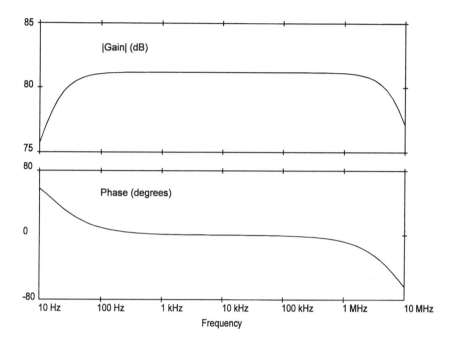

Figure 2.14. Gain (top) and phase (bottom) versus frequency for $R_F = 12$ KΩ.

As we have seen, small size arrays can be constructed using hybrid circuit technology with ceramic substrates. The size of the array is very important in quasi-diffused channels where the optical radiation from the array is collected by a lens.

Large-area photodiodes can be used to improve the optical power detected. However, due to the high price and capacitance associated with these devices, the detection area must be increased by using passive optical systems.

We have considered the basics of driver and preamplifier circuit design for wireless optical communications. Series driver and shunt driver are two attractive solutions for diffused channels where a LED array must be switched at low medium bit rates (up to 12.5 Mbps). Higher switching rates can be obtained in point to point applications using an emitter-follower drive circuit or emitter-coupled circuit. A wide dynamic range receiver preamplifier as the transimpedance configuration is required, however.

REFERENCES

[1] Adachi, S. "GaAs, AlAs, and Al$_x$Ga$_{1-x}$As: Material Parameters for Use in Research and Device Applications," *Journal of Applied Physics*, Vol. 58, No. 3, 1985, pp. R1–R29.

[2] Sze, S. M. *Physics of Semiconductor Devices*, 2d ed. New York: John Wiley & Sons, 1981.

[3] HAMAMATSU Technical Data, *High-Speed Infrared LEDs*, Japan, 1990.

[4] Antonetti, V. W., and R. E. Simons. "Bibliography of Heat Transfer in Electronic Equipment," *IEEE Transactions on Components, Hybrids, and Manufacturing Technology*, Vol. 8, No. 2, 1985, pp. 289–295.

[5] McPhillips, R. B. "Advanced Ceramic Materials for High Thermal Conductivity Substrate Applications," *Hybrid Circuit Technology*, August 1988, pp. 21–23.

[6] Dean, D. J. *Thermal Design of Electronic Circuit Boards and Packages*, Scotland: Electrochemical Publications, 1985.

[7] Wehrhahn, E. "Temperature Distribution on Substrates of Hybrid Circuits," *5th European Hybrid Microelectronics Conference*, Stresa, Italy, 1985, pp. 288–295.

[8] MOTOROLA Catalog. *Optoelectronics Devices*, DL118, REV3, Arizona, 1989.

[9] Gfeller, F. R., and U. Bapst. "Wireless In-House Data Communication via Diffuse Infrared Radiation," *Proceedings of the IEEE*, Vol. 67, No. 11, 1979, pp. 1474–1486.

[10] Koehler, D. "Semiconductor Switching at High Pulse Rates," *IEEE Spectrum*, Vol. 2, No. 11, 1965, pp. 50–66.

[11] Shumate, P. W., and M. DiDomenico. *Chapter 5, Lightwave Transmitters, Semiconductor Devices for Optical Communications*, Berlin: Springer-Verlag, 1980, pp. 161–200.

[12] Georgopoulos, C. J., and A. K. Kormakopoulos. "A 1Mbit/s IR LED Array Driver for Office Wireless Communication," *IEEE J. Solid-State Circuits*, Vol. 21, No. 4, 1986, pp. 582–584.

[13] Gabiola, F. J. "Contribución al estudio de técnicas de comunicaciones ópticas no guiadas para su aplicación en redes de datós," Ph.D. diss., Universidad Politécnica de Mádrid, Madrid, September 1992.

[14] Kozikowski, J. L. "Analysis and Design of Emitter Followers at High Frequencies," *IEEE Trans. Circ. Theory*, Vol. 11, No. 1, 1964, pp. 129–136.

[15] Gowar, J. *Optical Communication Systems*, London: Prentice Hall, 1984.

[16] RCA Electro Optics, *Short form catalog*, Canada, 1989.

[17] Muoi, T. V. "Receiver Design for High-Speed Optical-Fiber Systems," *Journal of Lightwave Technology*, Vol. 2, No. 3, 1984, pp. 243–267.

Chapter 3

Design of Optical Systems For IR Wireless Links

Juan C. Miñano

E.T.S.I. Telecomunicación
Ciudad Universitaria s/n. 28040 Madrid, Spain

3.1 INTRODUCTION

This section deals with the design of optical systems for transmitting incoherent radiation from an extended (nonpunctual) source to an extended receiver in such a way that the transmitted power is maximized with respect to certain constraints. The theoretical framework is that of geometrical optics and the method of approach is called nonimaging optics, which (unlike imaging optics) is a branch of optics interested in the transfer of the radiation and not in the image formation. The benefit of nonimaging optics is that it maximizes the power transfer. This is an important point that is treated in Chapter 5 (for outdoor IR links) and in Chapter 4 (for indoor IR links). The reader will find in this chapter a description of the latest method of design of nonimaging devices and the comparison of the devices resulting from this method with other nonimaging devices, and with conventional (imaging) systems from the point of view of radiation transfer. This chapter does not include a description of the methods of design of imaging devices. Such descriptions can be found in classical books on optics [1].

The first works in nonimaging optics appeared in the mid 60s under the signature of three different authors: Roland Winston in Chicago, Baranov in the USSR, and Ploke in Germany. Winston, whose work developed most of what is known today about nonimaging optics, started with a device whose purpose was to collect Cerenkov radiation. Years later, many more applications appeared: solar energy, astronomy, biology, sensors, and so forth (a detailed view of these applications can be found in [2]). In all of these applications the optical problem is related, partially or totally, to the transference of incoherent radiation from a source to a receiver. The rest of this section is devoted to discussing what the characteristics should be of the optimum optical systems for an IR point-to-point link.

Point-to-point links are characterized by the small angle of emission and reception. This case not only includes typical point-to-point links, but also those cases in which the radiation is received (or should be emitted) within a small angle, for instance, when the emitters and receivers of a network are seeing a small region of a wall where the radiation is reflected. Nonimaging optics may be applied to other cases where wide-angle emitters or receivers are used (for instance, it has been applied to static concentration for solar energy collection, or light pumping of lasers). The greatest benefits of optics when applied to aerial optical links are achieved when the radiation is within a small angle. For this reason, and for the sake of simplicity, only the point-to-point case will be considered here.

Consider the problem of transferring through the atmosphere the power emitted by an LED to a photodiode placed far away. Assume, for the task of simplicity, that the LED is a flat surface of area A_l, which emits the radiation homogeneously and isotropically (i.e., that the radiance R emitted by any point of the LED surface at any direction is constant). Assume also that the photodiode is a flat surface with an area A_r. If the losses through the atmosphere are not considered, and if the distance D between the LED and the photodiode is much greater than the LED and photodiode diameters, then the power received by the photodiode from the LED is $R A_l A_r \cos \beta_l \cos \beta_r / D^2$ where β_l and β_r are, respectively, the angles formed by the normal to the LED and to the photodiode, with the line linking the centers of both surfaces.

Adding an optical system to the LED and to the photodiode may increase the apparent size of those surfaces, and so may increase the power received by the photodiode from the LED. This apparent size will depend on the angle from which the optical system is being seen. For instance, assume that the optical system of the receiver is a thin lens placed at the focal distance f from a circular photodiode of radius p_r. If the optical losses are neglected, the apparent area of the photodiode when observed from a distant point along the optical axis will be the area of the aperture of the optical system. If the point of observation is also far from the receiver but the rays linking that point with the lens surface form angles greater than $\tan^{-1}(p_r/f)$ with the optical axis, then the apparent area of the photodiode will be zero.

In general, the apparent area of the photodiode is a function that depends on the distance from the receiver to the point of observation and on the angle formed by the optical axis with the line linking the point of observation with the center of the receiver's aperture. When the distance between the receiver and the point of observation is much greater than the aperture diameter, then the dependence of the apparent area on the distance disappears. Assuming that both emitter and receiver use optical systems and that the distance between them is big enough, the power received by the photodiode from the LED can be written as $R \, \mathbf{A}_l(\beta_l) \, \mathbf{A}_r(\beta_r) \cos \beta_l \cos \beta_r / D^2$, where $\mathbf{A}_l(\beta_l)$ and $\mathbf{A}_r(\beta_r)$ are respectively the apparent areas of the LED and photodiode.

The total power leaving the emitter is given by $\int R \, \mathbf{A}_l(\beta_l) \cos \beta_l \, d\Omega$ ($d\Omega$ is a differential of solid angle), which necessarily must be smaller than the total power emitted by the LED W_e. From this inequality we can derive a limitation for the possible functions $\mathbf{A}_l(\beta_l)$. In a similar way we will derive a limitation for the functions $\mathbf{A}_r(\beta_r)$; in fact, the conclusions we obtain for one of these functions must be applicable to the other function, since these functions depend only on the optical system and not on the nature of the LED or photodiode surfaces. To simplify the following reasoning, we shall only consider the emitter optical system.

Let us state the problem in the following way: assume that the optical system of the emitter should be designed to work optimally if the error of alignment is not greater than a certain angle θ_a, that is, this optical system should emit its radiation within a cone of angle θ_a around the optical axis in such a way that (1) a receiver placed far away from the emitter and perfectly pointed to it receives the same power no matter what the alignment error of the emitter (always that it is smaller than θ_a) is, and (2) all the power radiated by the LED W_e is sent within the cone of angle θ_a. These conditions for the optimum emitter are equivalent to saying that the radiation intensity of the emitter at a distant point is zero for angles greater than θ_a and constant otherwise, being that constant $W_e/\{2\pi(1 - \cos \theta_a)\}$ (observe that $2\pi(1 - \cos \theta_a)$ is the solid angle of the cone of angle θ_a). The radiation intensity is given by $(R \, \mathbf{A}_l(\beta_l) \cos \beta_l$ (the power emitted per stereoradian). The radiance R at the optical system aperture cannot be greater than the radiance at the LED surface and it is equal if there are no optical losses [2].

If θ_a is a small angle, then $\cos (\beta_l) \simeq 1$ when $\beta_l < \theta_a$ and the prior condition for the optimum emitter can be stated in simpler terms by requiring that $\mathbf{A}_l(\beta_l) = 0$ if $\beta_l < \theta_a$ and $\mathbf{A}_l(\beta_l) = $ constant otherwise. This constant is $W_e/(R \, \pi \sin^2\theta_a) = A_l/\sin^2\theta_a$ if all the power emitted by the LED is sent by the optical system within the cone of angle θ_a. The power emitted by the LED is $W_e = R \, \pi \, A_l$ if it is surrounded by air. If the LED is surrounded by a dielectric medium with refractive index n, then the power emitted by the LED is $W_e = R \, \pi \, A_l \, n^2$ (provided that n is not greater than the refractive index of the LED material and that the material surrounding the LED has an appropriate shape to avoid part of the energy being reflected back to the LED). Then, the apparent area of the emitter within the cone of angle θ_a will be $\mathbf{A}_l(\beta_l) = A_l \, n^2/ \sin^2\theta_a$ and the optical gain of the emitter with respect to the LED alone will be $n^2/ \sin^2\theta_a$.

Henceforth, only the case when θ_a is small ($\cos \theta_a \simeq 1$) will be considered. Summarizing for this case, the best we can do with respect to the function $\mathbf{A}_l(\beta_l)$ is that $\mathbf{A}_l(\beta_l) = A_l \, n^2/ \sin^2\theta_a$ if $\beta_l < \theta_a$ and $\mathbf{A}_l(\beta_l) = 0$ otherwise.

There is still something we may require from the optimum optical system. The aperture of the optical system must have an area which is at least equal to the maximum of $\mathbf{A}_l(\beta_l)$ (i.e., its minimum area is $A_l \, n^2/\sin^2\theta_a$). The additional requirement is that the aperture of the optical system equates this minimum. In a first approximation, the smaller the aperture of the optical system, the smaller its cost.

Using conventional imaging devices, it is very easy to achieve a stepped function $\mathbf{A}_i(\beta_l)$ such that $\mathbf{A}_i(\beta_l) = 0$ for $\beta_l > \theta_a$ and $\mathbf{A}_i(\beta_l)$ constant otherwise. Nevertheless, only using complex systems, or in some particular cases, it is possible to get close to the upper bound $\mathbf{A}_i(\beta_l) = A_l\, n^2/\sin^2\theta_a$ for $\beta_l < \theta_a$ [2]. For instance, an optical system formed by an LED placed at the focal plane of an $f/1$ lens transmits, within $\beta_l < \theta_a$, 20% of its total power. This will be expressed by saying that this device has a total transmission $T = 0.2$. If the f number increases up to 1.5, then the total power transmitted within the cone $\beta_l < \theta_a$ is only 10% ($T = 0.1$). The power transmitted within the cone may be increased up to 25% using a parabolic mirror. If two optical stages are used, then the power transmitted may be increased. For instance, if the bound of an LED surface is placed at the inner aplanatic surface of a refracting sphere (refractive index 1.5), and if the outer aplanatic surface coincides with the focal plane of a $f/1$ lens, then the power transmitted within $\beta_l < \theta_a$ may get close to 45% of the power emitted by the LED. Increasing T_t from this value using imaging devices use to be very complex. Nevertheless, the performance of imaging devices may be good enough for some applications; for instance, a device with $T = 0.1$ is equivalent to a $T = 1$ concentrator with a 10dB attenuator. In some cases this loss may be compensated for with extra amplification in the receiver.

Nonimaging devices usually can achieve values of T greater than 0.85. They have an additional advantage: nonimaging concentrators use to be very simple (no more than two optical elements), and some of them are very compact. Their principal disadvantage when compared with imaging devices is that their availability in the market is very limited.

Now we are going to consider the case of the receiver with more detail. Again, only the case of a link in which the emitter is far away from the receiver is considered. This is because, in these cases, the optimization of the optical systems for maximum transfer of power is more critical. Let us start with the problem of concentrating onto a photodiode the radiation coming from a source placed at infinity.

In order to state more properly the problem, consider that the source is radiating homogeneously (i.e., with constant radiance) inside a cone of rays, which form an angle smaller than θ_a with the z axis (see Fig. 3.1). Outside the cone, the source does not radiate. Assume that the active area of the photodiode is a circle of radius unity placed at the plane $z = 0$ and centered at $x = 0$, $y = 0$. The problem is stated as the design of an optical system casting onto the photodiode the maximum power coming from the source with the minimum entry-aperture area. The maximum power cast on the photodiode from a source of radiance R is [2] $A_r n^2 \pi R$ where A_r is the photodiode area, n is the refractive index of the medium surrounding the photodiode, and R is the radiance of the source. When the photodiode is receiving such power, all the points of its surface are being isotropically

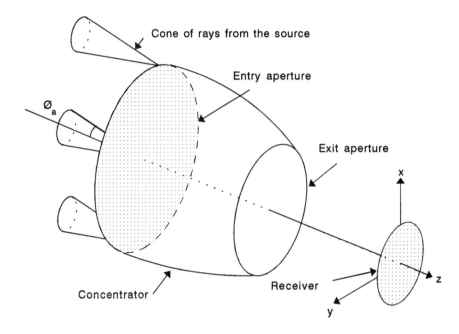

Figure 3.1 Scheme of a general concentrator.

illuminated by rays coming from the source directly or through an optical system. This upper bound of the power received by the photodiode is independent of the angular spread of the source. The condition of minimum entry-aperture area is imposed to maximize the collection efficiency of the optical system (i.e., to maximize the ratio of power directed onto the photodiode with respect to the power intercepted by the optical system). Maximizing this ratio is usually (but not always) related to the minimization of the cost of the optical system. The consideration of the cost in the design process is difficult because there are many factors affecting it.

Generally speaking, we can consider that the cost increases with the entry-aperture area, so minimizing this area is a goal of the design procedure given that the photodiode receives the maximum power from the source. The minimum entry-aperture area is that which intercepts an amount of power from the source equal to $A_r n^2 \pi R$. This area is $A_r n^2/\sin^2\theta_a$. Then, the function $\mathbf{A}_r(\beta_r)$ of the optimum optical system for the receiver is $\mathbf{A}_r(\beta_r) = A_r n^2/\sin^2\theta_a$ when $\beta_r < \theta_a$ and, so, the optical gain for this range of values of β_r is $n^2/\sin^2\theta_a$, as in the case of the emitter. If $\beta_r > \theta_a$, then $\mathbf{A}_r(\beta_r)$ is necessarily zero; otherwise, the power received by the photodiode would overcome the upper bound.

The case of source considered above not only comprises sources radiating with a certain angular spread, but also those cases in which the source can be considered as punctual and a certain misalignment between the source and the optical system associated to the photodiode must be allowed. In these last cases, the radiance of the equivalent extended source is smaller than that of the punctual source. An example of this case is the optical system used to concentrate solar energy on the surface of a photovoltaic cell (or any other type of receptor). The direct radiation coming from the sun has an angular spread of 0.26 degrees ($\theta_a = 0.26$ deg). Typically, the optical system is designed for an angular spread of 1 degree in order to collect part of the circumsolar radiation and to allow optical and suntracking errors.

It is clear now that from the optical point of view, the optical system of the receiver is qualitatively equivalent to the optical system of the emitter. When the distance between both optical systems is much greater than their aperture diameters, then the bundle of rays emitted (received) at the aperture of these optical systems has a relatively small angular spread when compared with the bundle of rays leaving (reaching to) the LED (the photodiode). The optimum optical systems must, in both cases, couple the bundle of rays at the entry aperture with the bundle of rays at the LED or photodiode.

A deeper analysis of the problem is done in the next section, and the keys for the optical design procedures are shown. Other sections are devoted to particular designs and their corresponding analysis. Henceforth the study is restricted to the optical system of the receiver, which used to be called concentrator because it concentrated on the photodiode the power received at its entry aperture. A new function $\mathbf{T}(\beta_r)$ (transmission-angle function) is used instead of $\mathbf{A}_r(\beta_r)$. The relationship between them is $\mathbf{A}_r(\beta_r) = C_g A_r \mathbf{T}(\beta_r)$ where C_g is the ratio of the entry aperture area to A_r (C_g is called geometrical concentration). Observe that $\mathbf{T}(\beta_r) < 1$.

3.2 STATEMENT OF THE PROBLEM

Let Σ_i be the entry aperture of the optical system and Σ_o be the exit aperture. Assume that both the entry and exit apertures lay on a $z =$ constant plane. This assumption simplifies the following reasoning and does not introduce a lack of generality. Let n_i be the index of refraction of the medium between the source and the entry aperture and n_o be the index of refraction of the medium between the exit aperture and the photodiode. Typically, $n_i = 1$ and n_o varies from ≈ 1 to 1.5. Let p be the optical direction cosine of a ray with respect the x axis. For instance, p is n_i times $\cos(\alpha)$ (α is the angle formed between the ray and the x axis) when p is calculated at a point of the trajectory of the ray between the source and the entry aperture, and p is $n_o \cos(\alpha)$ when the point of the trajectory where the calculations are done is between the exit aperture and the photodiode. Let q be the optical direction cosine of the ray with respect to the y axis and let r be the optical direction cosine with respect to the z axis. Then $p^2 + q^2 + r^2 = n^2$, n being the index of refraction of the point where p, q, and r are calculated.

Each ray reaching the entry aperture Σ_i can be characterized with four parameters. These parameters can be, for instance, the coordinates x, y of the point of interception of the ray with the entry aperture and the coordinates p, q of the ray at this point of interception. Similarly, the rays issuing from the exit aperture Σ_o can be characterized by another set of four parameters, for instance the coordinates x, y of the point of interception of the ray with the photodiode and the optical direction cosines p, q of the ray at that point of interception. The set of rays linking the source with the entry aperture is represented in the space x-y-p-q (this space is called the phase space) by a region including the points x, y, p, q such that x, y is the coordinate of a point of the entry aperture and $p^2 + q^2 \le n_i^2 \sin^2\theta_a$. Let us call this set of rays M_i. Similarly, M_o is a region of the phase space x-y-p-q whose points represent the rays linking Σ_o with the photodiode.

The purpose of this section is to design an optical system in which the rays represented by M_i are transformed into the rays represented by M_o, and vice versa. In this case the rays of M_i and M_o are the same, the only difference between M_i and M_o being that M_i groups the representation x, y, p, q of the rays at Σ_i and M_o groups the representation x, y, p, q of the same rays at the photodiode. This coincidence between the rays of M_i and M_o will be denoted by $M_i = M_o$.

We consider a particular case when M_o includes all the rays reaching the photodiode. In this case, if the optical system couples M_i and M_o ($M_i = M_o$), then the photodiode is illuminated by all possible rays reaching it with the radiance of the source, and so the power on it is the maximum one [2] (this maximum does not affect optical systems with frequency shift [3,4], which are not considered here).

Note that the requirement for the optical system is that $M_i = M_o$, and nothing is said about the particular transformation of each one of the rays; just that M_i and M_o represent the same bundle of rays at two different surfaces (at Σ_i and at Σ_o). In general, actual optical systems do not achieve this condition perfectly and therefore, when the bundle of rays M_c connecting the source with the photodiode through the optical system is represented in the phase space using the coordinates x-y-p-q at Σ_i, then the resulting region M_{ci} does not coincide with M_i in the general case. Obviously, M_{ci} must be a subset of M_i because the definition of M_c includes the rays connecting source and photodiode through the optical system, that is, the rays of M_c should cross the entry aperture (and also the exit aperture). Similarly, the region M_{co} representing M_c at the exit aperture Σ_o is a subset M_o (and in practical cases $M_{co} \ne M_o$).

If M_i and M_o have to be the same set of rays (i.e., $M_i = M_o$), then the first necessary condition to be fulfilled is that the phase space volume (or etendue $\mathcal{E}$) of M_i and of M_o be the same, that is,

$$\mathscr{E}(\boldsymbol{M}_i) \equiv \int_{\boldsymbol{M}_i} dxdydpdq = \mathscr{E}(\boldsymbol{M}_0) \equiv \int_{\boldsymbol{M}_o} dxdydpdq \qquad (3.1)$$

This condition derives from the conservation of phase space volume or conservation of etendue (this conservation is known in mechanics as the Liouville's theorem, which is also one of the Poincaré's invariant) [2].

There is another condition to be fulfilled if the optical system to be designed has to transform the rays of $\boldsymbol{M}_i$ into the rays of $\boldsymbol{M}_o$ (and vice versa). This condition, which is known as the edge ray principle [2] or the edge ray theorem [5,6,7], establishes that to obtain $\boldsymbol{M}_i = \boldsymbol{M}_o$, it is enough that $\partial \boldsymbol{M}_i = \partial \boldsymbol{M}_o$ where $\partial \boldsymbol{M}_i$ and $\partial \boldsymbol{M}_o$ are, respectively, the sets of rays represented by the points of the borders of the regions $\boldsymbol{M}_i$ and $\boldsymbol{M}_o$ in the phase space. In other words, it is enough that the optical system to be designed transforms the rays of $\partial \boldsymbol{M}_i$ into the rays of $\partial \boldsymbol{M}_o$, and vice versa. Again, there are no requirements about which ray of $\partial \boldsymbol{M}_i$ has to be linked with a given ray of $\partial \boldsymbol{M}_o$.

The regions $\boldsymbol{M}_i$ and $\boldsymbol{M}_o$ are four-dimensional (i.e., to distinguish one ray of $\boldsymbol{M}_i$ from another it is necessary to give four parameters). The regions $\partial \boldsymbol{M}_i$ and $\partial \boldsymbol{M}_o$ are three-dimensional, so the edge ray theorem reduces the complexity of the design of the optical system substantially. Nevertheless, the problem is still quite complex, and as a result there is only one design method for such devices [6]. Unfortunately, the optical devices resulting from this method use, in general, media with a given distribution of the index of refraction, which is very impractical.

The design problem is simplified if only the rays contained in a plane are considered (i.e., if we only consider two-dimensional optical systems). In this case the source, the photodiode, and the entry and exit apertures are lines in a plane; and so, the sets of rays $\boldsymbol{M}_i$ and $\boldsymbol{M}_o$ are two-dimensional (i.e., the rays belonging to one of those sets can be distinguished among themselves by giving two parameters). Assume that the entry and exit apertures are segments of $y = $ constant straight lines in an x-y plane (x and y are Cartesian coordinates); then the rays of $\boldsymbol{M}_i$ are uniquely characterized by giving the point x of interception of the ray with the entry aperture and the optical direction cosine p (with respect to the x axis) of the ray at this point of interception. In a similar way, each ray of $\boldsymbol{M}_o$ is characterized by a couple x, p. The edge ray theorem also holds for these two-dimensional optical systems; then, if the optical system has to achieve $\boldsymbol{M}_i = \boldsymbol{M}_o$ it is enough that it ensures $\partial \boldsymbol{M}_i = \partial \boldsymbol{M}_o$. These two sets of rays ($\partial \boldsymbol{M}_i$ and $\partial \boldsymbol{M}_o$) are now one-dimensional, and the problem of design is much more affordable.

There are four design procedures for these twodimensional optical systems. Two of them are explained in [3,8]. Another one can be found in [7,9]. The last design procedure is explained in detail in this chapter.

Actual optical systems are not two-dimensional in the sense that they are not contained in a plane, so the whole design does not stop when the two-dimensional optical system is obtained. After the two-dimensional design, the three-dimensional optical concentrator is typically generated by rotation (systems with rotational symmetry), or by translation (systems with linear symmetry) of the two-dimensional design. The first group of systems are used when the three-dimensional source and the receiver (the photodiode in our example) have rotational symmetry, and the second group is used when those surfaces have linear symmetry. The final three-dimensional design is not ideal in the sense that, in three-dimensional geometry, not all the rays of M_i are rays of M_o, and not all the rays of M_o are rays of M_i (i.e., not all the rays emanating from the source and intercepting the entry aperture are finally sent to the photodiode, and not all the rays linking the exit aperture and the photodiode come from the source). In the general case, the resulting three-dimensional optical systems adequately approach the condition $M_i = M_o$. Nevertheless, a ray tracing is needed to calculate it. This ray tracing will be the final step of the design if its result is satisfactory enough.

Henceforth we shall consider only two-dimensional optical systems unless three-dimensional is specified. Nevertheless, we shall use expressions such as entry-aperture area or receiver area even though the proper name for two-dimensional devices would be entry-aperture length or receiver length.

3.3 RELATIONSHIP BETWEEN THE CONCENTRATION AND THE ANGULAR SPREAD

The most important advantages of nonimaging devices appear when the receiver must have the smallest possible area conditioned to a given area of the entry aperture, and to cast to the receiver the radiation coming through M_i (i.e., the radiation coming from the source and intercepted by the entry aperture). This case used to be very important in many applications. For instance, if the receiver is a photodiode, a smaller area means a smaller capacitance, a higher cut-off frequency, and a smaller dark current (under the same technological conditions). The same case is found when the problem is stated in another two ways. The first is when it is required to have the greatest possible entry-aperture area conditioned to a given area of the photodiode and requiring that the radiation coming from M_i be directed to the receiver (this is obviously the same case as the one mentioned at the beginning of the paragraph). The greater the entry-aperture area, the greater the power on the photodiode. The second way to state the problem happens when the area of the photodiode and entry aperture are given and we want to design the optical device for the greatest possible angular spread of the source θ_a.

Let us call C_g the ratio of the entry-aperture area over the receiver area. The first question related to be solved is the relationship between the angular spread of the source, the entry-aperture area, and the receiver area. To simplify the following reasoning we shall restrict our analysis to sources placed at infinity radiating with an angular spread of $\pm \theta_a$ around the direction $p = 0$ (normal to the entry aperture), and to receivers and entry apertures placed at $y =$ constant lines in an x-y Cartesian plane. Assume that $n_i = 1$ (n_i is the refractive index of the medium between the source and the entry aperture), and $n_o = n$. Since we require that all the radiation coming from the source to the entry aperture be cast to the receiver, $M_i \subseteq M_c$. Since $M_c \subseteq M_i$ because of the definition of M_c, $M_c = M_i$ and so

$$\mathscr{E}(M_i) = \mathscr{E}(M_c) \tag{3.2}$$

The region M_i is formed by all the points x, p of the phase space where x belongs to the entry aperture (for example, $-A_e/2 \le x \le A_e/2$, where A_e is the entry-aperture length) and $-\sin \theta_a \le p \le \sin \theta_a$, so

$$\mathscr{E}(M_i) = 2A_e \sin \theta_a \tag{3.3}$$

because the etendue of a set of rays M in two-dimensional geometry is given by [2]:

$$\mathscr{E}(M) \equiv \int dxdp \tag{3.4}$$

The set of rays M_c is a subset of the set formed by all the rays reaching the receiver M_{om}, whose etendue is $\mathscr{E}(M_{om}) = 2 A_r n$ (observe that x varies in that case along a line of length A_r and that p varies from $-n$ to n). Then,

$$\mathscr{E}(M_c) \le 2A_r n \tag{3.5}$$

Using (3.2), (3.3), and (3.4), it is found that

$$C_g \le \frac{n}{\sin \theta_a} \tag{3.6}$$

The equation (3.6) establishes an upper bound for the ratio $C_g = A_e/A_c$, if $M_i \subseteq M_c$ and if the angular spread of the source is fixed. The upper bound is reached when $M_i = M_{om} = M_c$. The devices fulfilling such condition are called optimal concentrators.

Another upper bound is found when three-dimensional axisymmetrical devices are considered. In this case, the equivalent to (3.6) is [2]:

$$C_g \leq \frac{n^2}{\sin^2\theta_a} \tag{3.7}$$

Let R be the radiance (watts per square meter and per stereoradian) of the source. Then the irradiance on a receiver of a three-dimensional optimal concentrator with symmetry of revolution is $\pi n^2 R$, assuming that the optical losses (by reflection, absorption, or dispersion) are zero. This is the highest level of irradiance that can be found on the receiver with the source of radiance R. Nevertheless, to obtain such a level of irradiance on the receiver it is not necessary to use an optimal concentrator. It is enough that $M_c = M_{om}$. Assume that there is a nonoptimal concentrator such that $M_c = M_{om}$ and $M_c \neq M_i$. Since $M_c \subset M_i$ because of the definition of M_c, the entry-aperture area A_e of a nonoptimal concentrator producing the maximum irradiance on the receiver is necessarily greater than the A_e of an optimal concentrator, because $\mathscr{E}(M_c) < \mathscr{E}(M_i)$. So, we can conclude that an optimal concentrator is that producing the highest irradiance on the receiver using the minimum entry-aperture area. This conclusion is valid for two-dimensional and three-dimensional devices if the source is placed at infinity.

These types of optical devices (optimal concentrators) are of particular interest. They collect all the radiation received from the source at its entry aperture and send it to a receiver, which has the minimum necessary area to collect all this radiation. We can use a nonoptimal concentrator such that the average irradiance on the photodiode is as high as when the photodiode is coupled with an optimal concentrator but, necessarily, the entry aperture of the non optimal concentrator will be greater; or we can use a smaller photodiode than the one required in the optimal concentrator, but then some radiation arriving at the concentrator entry aperture from the source is not sent to the photodiode.

The next section is devoted to the design of an optimal concentrator in two-dimensional geometry formed by a combination of a lens and a reflective non-imaging concentrator, called flow-line concentrator (FLC) [10,11] or trumpet concentrator.

3.4 COMBINATION OF ASPHERIC NONIMAGING LENSES WITH THE FLC

The FLC is a three-dimensional concentrator formed by a reflector, which is obtained by rotating a branch of an hyperbola around the axis of symmetry not containing the foci. Then, the foci describe a circle, F, inside of which is another circle, R, described by the apex of the hyperbola. This last circle is the place of the receiver (see Fig. 3.2).

Let us call M_{i2} the set of rays crossing the entry aperture and being directed to the circle F. The hyperbolic reflector sends all the rays of M_{i2} to the circle R in

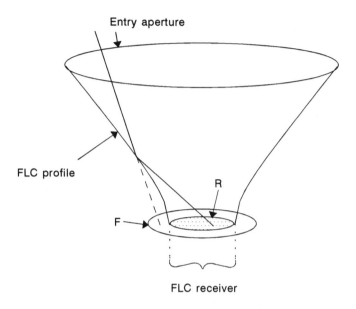

Figure 3.2. The FLC couples the rays reaching the circle R with the bundle of rays on the entry aperture in such directions as they would arrive at the circle F if no concentrator existed.

such a way that all the rays reaching the receiver (M_{om}) are rays of M_{i2}, so $M_{i2} = M_{om}$. The main purpose of this section is to design a lens that couples the set of rays coming from a source at infinity (M_i) to the set M_{i2}. Afterwards, the FLC will couple M_{i2} to M_{om}. The lens will be designed in two-dimensional geometry and then, a three-dimensional lens will be obtained by rotation. The three-dimensional lens would be analyzed with a ray-tracing procedure.

During the design of the lens we call receiver to the circle F, which is in fact the receiver of the rays sent by the lens. For the sake of generality, the source will be considered to be at a finite distance from the concentrator in the explanation of the design method.

In order to fix the conditions of the design let us assume that the receiver width is 2 (the receiver edges R and R' are at $x = -1$ and $x = 1$) and that the source is at the segment SS' as it is shown in Figure 3.3.

Figure 3.4 shows the representation of M_i and M_o in the phase space x-p.

It is well known that a single refractive surface can image sharply a bundle of rays into a point if each point of the surface is crossed by a single ray of the bundle. In general, a single refractive surface can transform a given bundle of rays into another one, which is predetermined if there is no more than one ray crossing each point of this surface. These refractive surfaces are called Cartesian ovals [12].

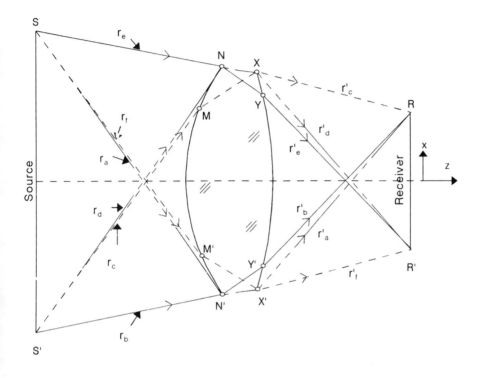

Figure 3.3 Location of the source and receiver and representation of some edge rays. Rays having the same subindex are the same ray.

Our problem is slightly different: there are two surfaces to be designed (the two surfaces of the lens) and each point of the two surfaces is crossed by two edge rays (except the extreme points of these surfaces which are crossed by a bundle of edge rays). The solution to this problem can be obtained using a "point-by-point method" similar to the one used by Schulz in the design of aspheric lenses [13,14].

Prior to applying this method we shall impose certain conditions on the transformation of the rays of ∂M_i into the rays of ∂M_o. These conditions derive from the statement of the problem. For instance, note that the rays reaching at the extreme point of the lens N (or N') cannot be the same as the rays departing from the extreme point of the lens X (or X') unless the lens has zero thickness at its edges. Only the ray r_a (and its symmetric counterpart r_d) crosses N' and X' (r_d crosses N and X, see Fig. 3.3): the trajectory of r_a reaches the point N' of the entry aperture with the most negative value of p (this ray comes from S). This ray must cross the most x-negative point of the exit aperture (point X') and the value of p of this ray at the point X' must be the highest one when compared to the other

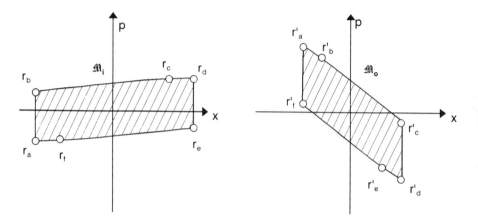

Figure 3.4. Representation in the phase space of M_i (left) and M_o (right). Some special edge rays are marked with a dot and their trajectory can be seen in Figure 3.3.

rays of ∂M_o crossing X'. Then the x-p representation of the ray r_a at the exit aperture must be r (i.e., the ray linking the x-negative edge of the exit aperture and the x-positive edge of the receiver).

Because of the symmetry of the lens, the conditions are only stated for the rays crossing the x-positive side of the lens (see Fig. 3.4; the notations r and r' mean the same ray before and after crossing the lens). The conditions are (1) the ray r_d (a corner of ∂M_i) is transformed into the ray r (of ∂M_o); (2) the rays of the other corner of ∂_i, that is, r_e is transformed in a ray (r) that crosses the lens exit aperture at a point Y different from X; (3) the other ray of the corner of ∂M_o, r come from the ray r_c that crosses the entry aperture at a point M different from N. Similar conditions hold for the rays r_a, r_b, and r_f.

The above conditions determine the portions MN (and $M'N'$) and XY (and $X'Y'$) of the two surfaces of the lens: the profile MN is a portion of a Cartesian oval that images the rays coming from S' (between r_c and r_d) at the point X; the profile YX is also a portion of a Cartesian oval that images N at R'.

The points R and R' are assumed to have coordinates $x = 1$ and $x = -1$, respectively. The size and position of the source (relative to the receiver) are assumed to be known. The design procedure is as follows. First, the etendue $\mathscr{E}$ of the manifold M_i is chosen. After this selection, the points N and X are chosen arbitrarily, bearing in mind that the etendue $\mathscr{E}$ of the manifolds M_i and M_o is the same. This means that point N must lay on the hyperbola $|NS'| - |NS| = \mathscr{E}/2$ and the point X must lay on the hyperbola defined by $|XR'| - |XR| = \mathscr{E}/2$, where $|XR|$ means optical length from X to R (see [2] or the calculation of the etendues). The selection of N and X fully determines the Cartesian ovals MN and

XY. The first Cartesian oval is the one crossing N and imaging S' and X. The second one is a Cartesian oval crossing X and imaging N and R'. Point M is the intersection of the ray r_c coming from R and crossing X (note that the normal to the refractive surface at X is known since the Cartesian oval crossing X is known, so it is possible to trace the trajectory of the ray r_c inside the lens and then calculate the point M). In a similar way the point Y can be calculated with the ray r_e coming from S and refracting at N.

Now consider an arbitrary point O of the lens surface between M and N (see Fig. 3.5). The ray r_g impinging on O from S must be directed to R'. The trajectory of r_g inside the lens can be easily calculated (with the refraction law) since the profile MN is known. Then, point P, where the ray r_g leaves the lens, can be calculated by imposing that the optical length l from S to R' be the same through r_e or through r_g, that is, $|SNYR'| = |SOPR'|$ (note that $|SNYR'|$ can be calculated because all the points S, N, Y, and R' are known). A new portion of the rightmost surface of the lens can be obtained by applying the above procedure to all the points between N and M. The derivative of this new portion of surface at the

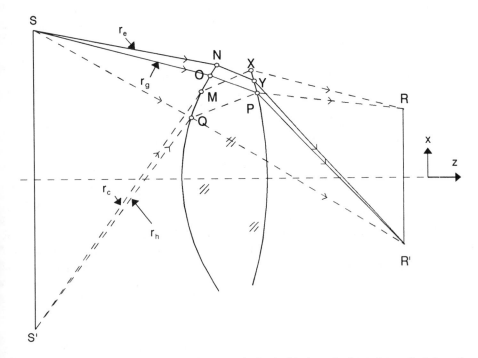

Figure 3.5 The remaining points of the lens are obtained with the point by point method departing from the Cartesian ovals *NM* and *XY*.

point P can be calculated using its neighboring points or by the application of the refraction law to ray r_g at point P.

Now consider ray r_h that links P and R. The trajectory of r_h inside the lens can be calculated because the normal to the profile at P is known. This ray must impinge on the leftmost surface of the lens at a point Q in such a way that r_h comes from S'. If the lens is symmetric (with respect to the z axis), the optical length along r_h from S' to R the must be l, and so the point Q and the normal to the profile at Q can be calculated as we did with the point P. A new portion of the left-most surface of the lens can be calculated by repeating the preceding procedure with the rays linking R with the calculated portion of the right-most surface of the lens (i.e., the portion XY and the portion calculated during the first step of the procedure). The remaining points of the lens are calculated by repeating the procedure.

The x-positive side of the two lens profiles is calculated according the above procedure. The other side of the lens is obtained by symmetry. Generally, the lens obtained with this procedure is not normal to the z axis at $x = 0$ (this is not a necessary condition for the design of the lens), and so a discontinuity of the derivative of the profile may exist. To get lens profiles normal to the z axis at $x = 0$, it is necessary to iterate the design procedure with different initial points N and X. First, point N can be kept in its initial position and point X is moved along the hyperbola $|XR'| - |XR| = \mathscr{C}/2$ until the left-most surface of the lens is normal to the z axis at $x = 0$ (more than a single solution can exist). Second, the point X is kept constant and the point N is moved along the hyperbola $|NS'| - |NS| = \mathscr{C}/2$ until the right-most surface of the lens is normal to the z axis at $x = 0$. By iteration of this procedure it is possible to find a lens having both surfaces normal to the z axis at $x = 0$. Finally, when the x-positive side of the lens has been designed with surfaces normal to the z axis at $x = 0$, the x negative side of the lens is obtained by symmetry.

Generally, there is not a single solution. To choose the best one it is possible to consider other features of the lenses. For instance, their performance as three-dimensional lenses, the thickness at the center of the lens, and so forth. The thinnest lens is that where N coincides with X.

This design procedure ensures that the manifolds ∂M_i and ∂M_o (see Fig. 3.4) are the same except for a small subset of rays crossing the central region of the lens. In order to study this small subset of rays, assume that the x-positive side of the lens is designed so that both surfaces of the lens are normal to the z-axis at $x = 0$. The design procedure ensures that all the rays of ∂M_i impinging on the entry aperture from the point N to L (see Fig. 3.6) are rays of ∂M_o after crossing the lens.

The same can be said of the rays that cross the lens through the portion XZ of the right-most surface of the lens. The design also ensures that the rays of ∂M_i coming from S' and impinging on the portion C_lL of left-most surface of the lens

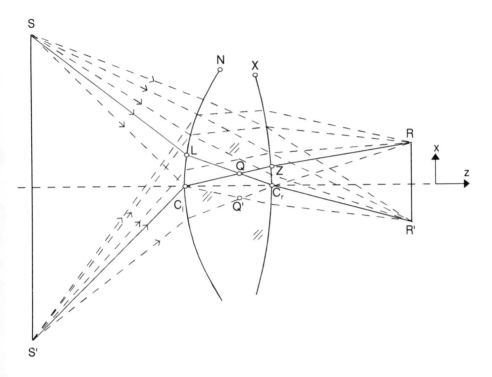

Figure 3.6. Construction of the lens at the center where the method requires an additional degree of freedom to rigorously obtain an ideal nonimaging concentrator.

become rays of ∂M_o (in the case shown in Fig. 3.6, these rays are focused at R), and that the rays crossing $C_r Z$ (right-most surface) and focused to R' are rays of ∂M_i coming from S. Let us construct the x-negative side of the lens by symmetry with respect to the z axis. The design method does not ensure that the rays from S impinging on the lens through $C_l L$ (these rays belong to ∂M_i) will be imaged at R', with the exception of two of these rays: those crossing the points C_l and L (see Fig. 3.7). The same situation can be viewed at the exit aperture of the lens: there is no evidence that the rays focused to R from $C_r Z$ are rays coming from S', excepting the rays crossing the points C_r and Z. The portions $C_l L$ and $C_r Z$ are fully determined by the design procedure, so there are no more degrees of freedom to solve this problem unless we accept, for example, the existence of an additional small lens (with different refractive index) between Q and Q'.

In practice, this problem does not exist and the rays impinging on the lens through $C_l L$ and coming from S result in being focused on R' up to the accuracy of the ray-tracing program (a similar thing happens with the rays focused on R and

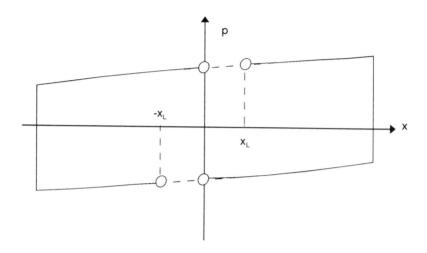

Figure 3.7. The method ensures that the part of ∂M_i represented with a solid line in this figure is transformed in rays of ∂M_o.

crossing C, Z), so $\partial M_i = \partial M_o$ in two-dimensional geometry. Nevertheless, we cannot establish rigorously that the method ensures $\partial M_i = \partial M_o$. Since this fact has no practical importance, we shall proceed as if ∂M_i and ∂M_o were absolutely the same manifold of rays (in two-dimensional geometry). The edge ray theorem establishes that $\partial M_i = \partial M_o$ is a sufficient condition to get $M_i = M_o$.

In any case, it would not be critical even if $\partial M_i \neq \partial M_o$ at the central regions of the profiles, because these central regions do not generate an important amount of area of the three-dimensional concentrators when these concentrators are generated by rotational symmetry around the z axis.

The x-negative side of the lens can also be generated using the same procedure with which the x-positive side has been calculated (i.e., following the procedure without stopping at $x = 0$). In this case the lens is, in general, asymmetric (excepting when the lens surfaces are normal to the z axis at $x = 0$) and so a three-dimensional rotational symmetric lens (with rotational symmetry around the z axis) cannot be constructed. Another possibility is to construct the x-negative side of the lens by symmetry even if the surfaces are not normal to the z axis at $x = 0$ and, so, accepting that the lenses have a kink at the center. This possibility has not been studied.

These lenses are also called *RR* concentrators (RRc), where *RR* means that the rays suffer two refractions from the source to the receiver. We shall keep this notation to avoid confusion with other concentrators. Hereafter the analysis is restricted to nonimaging lenses designed for a source placed at the infinity. Therefore, the source is not characterized by the position of the points S and S', but by the angular spread of the source $\pm \theta_a$.

3.5 THREE-DIMENSIONAL RAY TRACING OF THE *RR* CONCENTRATORS

As a part of the three-dimensional ray tracing of the lenses (or RRc) a two-dimensional ray tracing has been done to verify that $M_i = M_o$ in two-dimensional geometry (i.e., to verify that (1) any ray impinging on the entry aperture of the lens NN' with $-\sin(\theta_a) \leq p \leq \sin(\theta_a)$ reaches the receiver RR', and (2) any ray linking the exit aperture of the lens XX' and the receiver RR' has crossed the entry aperture NN' with $-\sin(\theta_a) \leq p \leq \sin(\theta_a)$.

Figure 3.8 shows the results of the three-dimensional ray-tracing analysis (no optical losses have been considered). The curves in this figure are the transmission-

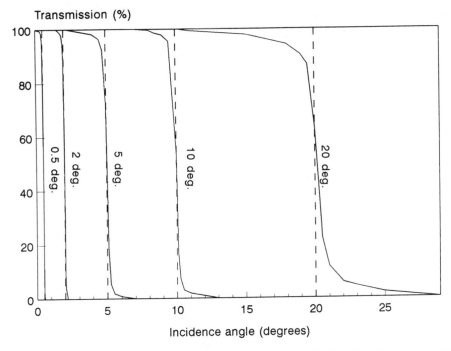

Figure 3.8 Transmission-angle curves for several three-dimensional RRc's designed for sources at the infinity. The number by each curve is the acceptance angle. (For other characteristics see Table 3.1.)

angle curves $T(\theta, \theta_a)$ for several nonimaging three-dimensional RR concentrators obtained by rotational symmetry (around the z axis) from two-dimensional RR concentrators with different acceptance angles θ_a designed according to the preceding section (other characteristics of these lenses are given in Table 3.1). These curves give the ratio of power transmitted to the receiver over the power transported by rays impinging on the entry aperture at a chosen incidence angle θ (assuming that these rays have a constant radiance). A total of 9,000 rays were traced for each incidence angle θ. The function $T(\theta, \theta_a)$ was more deeply explored near the transition (around $\theta = v_a$) to ensure that $T(\theta_i, \theta_a) - T(\theta_{i+1}, \theta_a) < 0.1$ (θ_i and θ_{i+1} are two consecutive values of θ where the function $T(\theta_i, \theta_a)$ is calculated).

If only meridian rays were considered, the transmission-angle curves would be $T(\theta, \theta_a) = 1$ if $\theta \leq \theta_a$ and $T(\theta, \theta_a) = 0$ if $\theta > \theta_a$, because the method of design of the RRc (RR concentrators). Nevertheless, these concentrators are not ideal in three-dimensional geometry (some skew rays with $\theta \leq \theta_a$ are not sent to the receiver and some other skew rays with $\theta > \theta_a$ are sent to it) so the transition from $T(\theta, \theta_a) = 1$ to $T(\theta, \theta_a) = 0$ is not abrupt in a three-dimensional RRc. In other words, the set of rays impinging on the concentrators entry aperture with $\theta \leq \theta_a$ (the set M_i) does not coincide with the set of rays linking the concentrator exit aperture and the receiver (M_o) in three-dimensional geometry. Nevertheless the etendue of these two sets $\mathscr{E}_{3D}$ is the same because of the method used for the construction of the concentrator. This etendue can be calculated at the entry or at the exit apertures [2],

$$\mathscr{E}_{3D} = \pi A_e \sin^2\theta_a = \frac{\pi^2}{4} (|XR'| - |XR|)^2 \tag{3.8}$$

where A_e is the area of the concentrator's entry aperture ($A_e = \pi x_N^2$, x_N being the coordinate x of the point N).

Table 3.1
Geometrical Characteristics and 3D Ray-Tracing Results of Some Selected RR Concentrators

Acceptance Angle, θ_a (degrees)	0.5	2	5	10	20
Geometrical concentration, C_{g3D}	3,600	225	36	12.25	2.25
Total transmission, $\mathbf{T}(\theta_a)$ (%)	97.5	98.0	96.9	96.6	95.9
Cutoff angular spread $\Delta\theta$ (degrees)	0.04	0.065	0.5	0.6	2
Estimation of the error, Err (%)	0.21	−0.20	−0.19	0.47	−0.40
Thickness at the center	101.5	25.65	9.5	3.99	1.33
Length/entry-aperture diameter, f	1.161	1.16	1.161	1.036	1.073
Exit aperture radius, R_o	20	5	2.5	2.5	1.2
Exit aperture to receiver distance, $x_R - z_X$	32.53	8.09	3.98	3.16	1.81
Entry aperture to receiver distance, $z_R - z_N$	72.35	17.95	7.84	4.28	2.27

Note: Lens refractive index = 1.483, receiver radius = 1.

The fact that $M_i \neq M_o$ does not mean that the edge ray theorem fails in three-dimensional geometry, because the method of design does not ensure $\partial M_i = \partial M_o$ in three-dimensional geometry (this is only ensured in two-dimensional geometry). Let M_{ci} be the representation of M_c at the entry aperture and let M_{co} be the representation of M_c at the exit aperture (remember that M_c is the set of rays linking the source and the receiver through the concentrator). The inverse edge theorem in three-dimensional geometry states that, since $M_{ci} = M_{co}$, then $\partial M_{ci} = \partial M_{co}$. Nevertheless, in general, $\partial M_{ci} \neq \partial M_i$ and $\partial M_{co} \neq \partial M_o$. So, it can occur that some rays of ∂M_i enter in the concentrator and are turned back (as it occurs in another concentrator, the compound parabolic concentrator, CPC), and this does not invalidate the edge ray theorem because the method of design does not ensure that $\partial M_i = \partial M_o$ in three-dimensional geometry (neither the method of design presented here, nor the method of design of the three-dimensional CPC with rotational symmetry).

An important figure characterizing the transmission of the concentrators is the total transmission $\mathbf{T}(\theta_a)$, which can be defined as the ratio of the etendue of M_c to $\mathcal{E}_{3D}$ (therefore $\mathbf{T}(\theta_a) \leq 1$). $\mathbf{T}(\theta_a)$ is the total flux transmitted inside the design collecting angle [2]. Its expression is:

$$T(\theta_a) = \frac{A_e \, \pi \int_0^{\theta_a} T(\theta, \, \theta_a) \sin 2\theta \, d\theta}{\mathcal{E}_{3D}} \tag{3.9}$$

$\mathbf{T}(\theta_a)$ is a quality factor of the concentrator (an ideal three-dimensional concentrator would have a rectangular cutoff at $\theta = \theta_a$, and so $\mathbf{T}(\theta_a) = 1$).

Another figure characterizing the transmission-angle curves is the angle difference $\Delta\theta = \theta_9 - \theta_1$, where θ_9 and θ_1 fulfill $T(\theta_9, \theta_a) = 0.9$ and $T(\theta_1, \theta_a) = 0.1$.

Table 3.1 shows the total transmission and $\Delta\theta$ obtained from the three-dimensional ray tracing of the RRc whose transmission-angle curve is shown in Figure 3.8. The table also gives some other features of the RRc, such as the geometrical concentration C_{g3D} (ratio of the entry and exit aperture areas), the ratio f of the length of the concentrator (difference between the z coordinates of the receiver and the center of the left-most surface of the lens) to the lens diameter, the thickness at the center of the lens (thickness), the exit aperture radius R_o, and the z coordinate of the points X and N relative to the receiver plane RR' (z_R-z_X and z_R-z_N respectively). The receiver radius is always 1 and the refractive index of the lens is 1.483.

To evaluate the error carried during the calculation of $\mathbf{T}(\theta_a)$ we have also calculated the following variable, Err

$$Err = 100 \times \left[1 - \frac{A_e}{\mathcal{E}_{3D}} \, \pi \int_0^{\pi/2} T(\theta, \, \theta_a) \sin 2\theta \, d\theta \right] \tag{3.10}$$

The integral in (3.10) times πA_e is the etendue of the set of rays crossing the entry aperture and reaching the receiver. This manifold coincides with M_o (this coincidence will not occur if some reverse ray departing from the receiver suffer a total internal reflection at the entry aperture surface). Since the etendue of M_o can be directly calculated (3.8), the comparison of both etendues can give a measure of the errors due to the ray tracing and to the numerical calculation.

Figure 3.9 shows the cross-section of an RRc with $\theta_a = 10$ deg (see Table 3.1 for other data) and the position of the points N, M, X, Y, L, and Z. When the acceptance angle of design θ_a is small and the lens to receiver distance is greater than twice (approximately) the lens diameter, then the RRc is approximately equivalent to a conventional imaging lens.

If an FLC is coupled to an RRc such that the entry aperture and the circle generated by the hyperbola foci of the FLC coincides respectively with the exit aperture and the receiver of the RRc, then a new concentrator is obtained. The receiver of the new concentrator is the receiver of the FLC (i.e., the circle generated

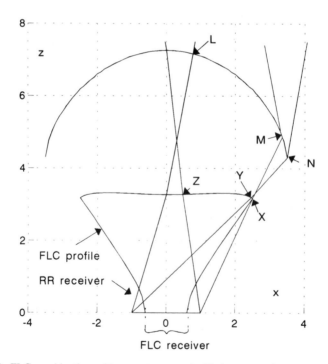

Figure 3.9 RRc-FLC combination with acceptance angle 10 degrees and geometrical concentration 33.2. The portion of Cartesian ovals (*MN* and *XY*) and the points *L* and *Z* are also shown.

by the apex of the hyperbola, which is smaller than the receiver of the RRc). This new concentrator has the same total transmission and angle transmission curves as the RRc, since the FLC is ideal in three-dimensional geometry (as we mentioned before) in the sense that all the radiation directed to the virtual receiver by the RRc is concentrated on the receiver of the FLC. The geometrical concentration (ratio of entry-aperture area to receiver area) reaches the upper bound established in (3.7) (the index of refraction n is equal to 1 in this case), so this new concentrator would be optimal if $T(\theta_a) = 1$. Notice in Table 3.1 that the achieved values of $T(\theta_a) = 1$ for the RRc-trumpet combinations are very close to 1. The total transmission increases when the acceptance angle decreases and the length from the receiver to the lens apex increases (i.e., when the RRc approaches a thin lens). Such combinations (thin lens-FLC) are detailed analyzed in [15].

3.6 THE XR CONCENTRATOR

There are other possibilities of designing two-dimensional optimal concentrators with the method explained in the preceding section. Generally speaking, the method requires a minimum of two optically active surfaces to be designed. The two surfaces provide a sufficient degree of freedom to accomplish the design. These two surfaces cannot both be refractive as in the case of the RRc; one can be refractive and another reflective, or both can be reflective. The practical designs will introduce additional restrictions, which can make some of these possibilities impractical. In this section we shall study the XR concentrator, a concentrator formed by a reflective (X) and a refractive (R) surface such that the rays coming from the source first intercept the reflective surface and, afterwards, the refractive surface. The medium between the concentrator and the source is assumed to have a refractive index of 1, so the receiver is immersed in a medium of refractive index $n > 1$ (note that the rays intercept the refractive surface only once) and the maximum achievable geometrical concentration C_g is increased by a factor n^2 (according to (3.7)) with respect to what was obtained with the RRc-trumpet combinations.

The design procedure is qualitatively identical to that of the RRc. The only difference is that now there is *one* reflective and *one* refractive surface instead of the *two* refractive surfaces of the RRc. Figure 3.10 shows one of these concentrators, designed for maximal concentration and for a source placed at infinity.

We shall describe the general procedure of design when the source is placed at a finite distance from the concentrator and when the set of rays M_o does not coincide with M_{om} (see Fig. 3.11). The location of the points N and X is done as before (i.e., taking into account that the conservation of etendue theorem has to be fulfilled). This means, in this case, that the point X must lay on the hyperbola $|XR'| - |XR| = \mathscr{E}/2$ and the point N on the hyperbola $|NS'| - |NS| = \mathscr{E}/2$ (the

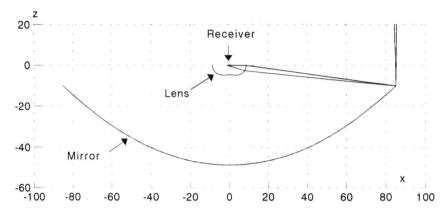

Figure 3.10 XRc for a source at the infinity subtending an angle of 1 deg with the *z* axis. This concentrator has maximal concentration. The Cartesian oval *XY* and the points *Z* and *L* are also shown.

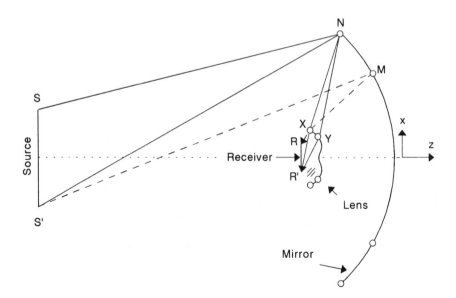

Figure 3.11 The construction of the XRc begins at the extreme points of the lens *X* and the mirror *N*.

points R and R' are assumed to be on the straight line $z = 0$ with $x = 1$). Note that the optical path lengths $|XR'|$ and $|XR|$ are the lengths between the corresponding point times the refractive index n.

Once the points N and X have been chosen, bearing in mind the above condition, the design of the profiles can start with the portion XY of the lens. This portion is a Cartesian oval that images the points N and R'. Since the position of the points N, X, and R' is known, such a Cartesian oval can be constructed easily. The point Y is obtained as the intersection of the Cartesian oval and the ray coming from S and reflecting at N.

The portion NM of the mirror is obtained in a similar way: NM is a part of an ellipse imaging the points S' and X. The point M is the intersection of this ellipse and the ray departing from R and crossing X. Observe the qualitative similarity between the portions XY and NM of the XRc and those portions of the RRc (see Fig. 3.3 and Fig. 3.11).

The design can now start from the portion XY. First, rays crossing XY are traced from R and a portion of mirror is calculated, requiring that these rays be the reflection of those reaching the mirror from the point S'. Second, rays coming from S are reflected in the last calculated portion of the mirror and another portion of the lens surface is obtained by requiring that these rays be focused at R'. The procedure is repeated until the profiles at $x = 0$ are known. If it is desired to have surfaces normal to the z axis at $x = 0$, then it is usually necessary to iterate all of the procedure with different starting points X and N until the surfaces are normal to the z axis at $x = 0$.

The design procedure does not always produce a real concentrator. Sometimes the method of design produces loops in the reflective or refractive surfaces caused by the caustics of the edge rays, which can be caused by an inappropriate selection of the points X and N. Whether the selection is proper or not is something that can only be determined during the design. In some cases these problems can be avoided by defining a virtual exit aperture, which is not touching the refractive surface. Note that it has been implicitly assumed that the exit aperture of the concentrator is the segment NN' (i.e., the exit aperture edges are on the refractive surface). This assignment is quite arbitrary and sometimes it can be useful to define an exit aperture that has no points in common with the refractive surface, for instance an exit aperture that is fully inside the dielectric medium.

As in the case of the RRc, there is a region around the center of the optical system (from point L to the center of the mirror and from point Z to the center of the dielectric, see Fig. 3.10) where there are not enough degrees of freedom to ensure that all the edge rays are correctly directed. Again, it is found that, up to accuracy of the two-dimensional ray tracing, these edge rays are correctly directed so $\partial M_i = \partial M_o$. Nevertheless, we cannot establish rigorously that this last condition is exactly fulfilled by a subset of the edge rays crossing the above-mentioned regions of the mirror and the lens.

3.7 THREE-DIMENSIONAL RAY TRACING OF THE XR CONCENTRATOR

The three-dimensional concentrators to be analyzed here are constructed by rotational symmetry around the z axis. The calculations are restricted to the case in which the source is at the infinity (as it was done before). Then, any ray of the source form an angle θ with the z axis such that $\theta \le \theta_a$ when they reach the mirror. The calculations are also restricted to the case of isotropic illumination of all the points of the receiver ($M_o = M_{om}$), that is, the point X is aligned with R and R'. The points N and M of the mirror become the same point in this case. The three-dimensional etendue of the manifold of rays crossing the entry aperture and reaching the receiver is $\mathscr{E}_{3D} = \pi n^2 A_r$, where A_r is the receiver area.

Figure 3.12 shows the transmission angle curves of several XRc's, one of which is the concentrator shown in Fig. 3.10 (the one corresponding to $\theta_a = 1$ deg). Other data on these concentrators appear in Table 3.2. For large values of θ_a (greater than 10–15 deg), the design method fails or the resulting portion of dielectric is so big that its shadow on the mirror makes the concentrator useless.

Table 3.2 includes two values of the total transmission $\mathbf{T}(\theta_a)$. The first one takes into account the shadow made by the dielectric on the mirror, that is, it is assumed that the light impinging on the entry aperture at the back side of the dielectric (or the back side of the receiver) is lost. The second one does not consider the shadow.

The ratio Err is calculated according (3.10), bearing in mind that $\mathscr{E}_{3D} = \pi n^2 A_r$. The dielectric thickness is measured at the center. z_N-z_R is the distance from the plane of the entry aperture to the plane of the receiver. This distance can be negative when θ_a is small, meaning that the dielectric and the receiver are fully inside the volume enclosed by the mirror surface and the plane of the entry aperture.

3.8 THE RXI CONCENTRATOR

In this concentrator, the rays suffer a refraction (R), a reflection (X) and a total internal reflection (I) in their trajectory from the source to the receiver. The RXIc is formed by two surfaces (see Fig. 3.13).

The lower surface is a reflector and the upper one acts as a lens when the rays come from the source to the RXIc, and as a reflector (by total internal reflection) when the rays have been reflected by the upper surface. The method of design of this concentrator starts by considering that the upper surface is different for the rays to be refracted than for the rays to be reflected. An upper surface for the rays to be refracted is proposed; then the upper surface for the rays to be reflected and the lower surface are calculated by a procedure similar to the design of the XRc. In a second step of the design procedure, the upper surface for the

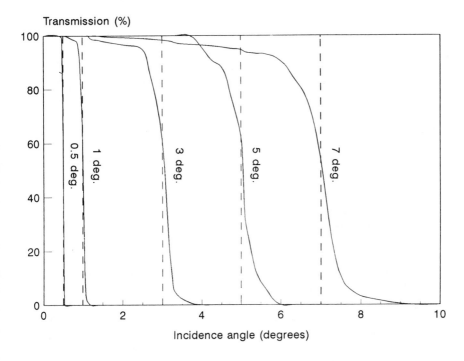

Figure 3.12 Transmission-angle curves for several three-dimensional XRc designed for sources at infinity with an angle θ_a. The curve numbering is the acceptance angle. (See Table 3.2.)

Table 3.2
Geometrical Characteristics and 3D Ray-Tracing Results of Some Selected XR Concentrators

θ_a (degrees)	0.5	1	3	5	7
Geometrical concentration, C_{g3D}	28880	7221	802.9	289.5	148
Total transmission, $\mathbf{T}(\theta_a)$ (%)	94.43	94.62	90.22	90.03	80.74
Total transmission without shadow, $\mathbf{T}(\theta_a)$	98.67	95.64	92.38	93.9	88.18
Cutoff angular spread, $\Delta\theta$ (deg)	0.05	0.15	0.65	1.0	1.5
Estimation of the error, Err (%)	0.26	−0.10	0.27	0.20	0.20
Length/entry-aperture diameter, f	0.257	0.287	0.313	0.356	0.348
Dielectric thickness at the center	13.56	4.56	2.57	2.57	2.33
Dielectric radius	35	8.5	4	3	3
Entry aperture to receiver distance, $z_N - z_R$	−5.42	10.08	5.88	5.91	3.77

Note: refractive index = 1.483, receiver radius = 1.

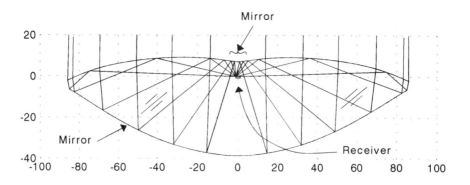

Figure 3.13 RXI concentrator with maximum concentration (C_{g3D} = 7,387, θ_a = $\pm 1°$ and n = 1.5) showing the trajectories of some edge rays.

rays to be reflected is proposed as the upper surface for the rays to be refracted. This step is repeated until both upper surfaces coincide. There is a central region of the upper surface which has to be metallized because there is not total internal reflection there. This region introduces a shadow for the rays coming from the source (in the case of the RXIc of Fig. 3.13, the area of the metallized region is 1%).

Figure 3.14 shows the angle-transmission curves for several RXIs with source at the infinity and with different acceptance angles. These curves do not take into account the shadowing of the upper metallized area. This shadowing is taken into account in the calculations of the total transmission, which is the number associated with each curve. The geometrical concentrations are those given by the upper bound in (3.7) for n = 1.5, that is, the C_g of the RXI of Figure 3.13 (θ_a = 1 deg) is C_g = 7,387.

3.9 COMPARISON OF THE XR AND THE RXI CONCENTRATOR WITH OTHER NONIMAGING CONCENTRATORS AND WITH IMAGE-FORMING SYSTEMS

The fundamental advantage of nonimaging concentrators versus imaging ones is that they typically use no more than two optical components to get concentrations close (in general greater than 90%) to the thermodynamic limit (3.7), while imaging concentrators need many more, particularly for small acceptance angles [2]. From another point of view, with the same number of optical components the nonimaging concentrators get much higher concentrations. For instance, a single parabolic mirror of revolution does not achieve concentrations greater than 25% of the limit

Transmission (%)

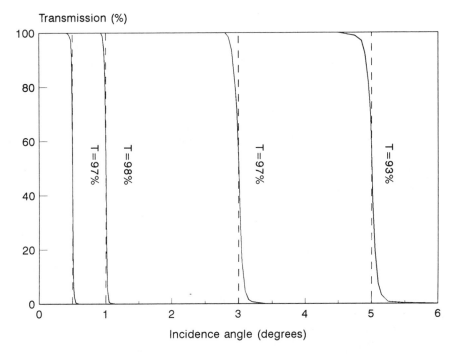

Figure 3.14 Transmission-angle curves for several three-dimensional RXIc for sources at the infinity with angle θ_a. The transmission includes the shadowing of the upper mirror.

[2], and not greater than 56% of the limit when the parabolic mirror is combined with a spherical refracting surface with $n = 1.5$ [16]. Increasing the number of components improves the concentration in an image-forming device, but only concentration levels as high as in the nonimaging concentrations are achieved, except in some particular cases with systems or materials that used to be impractical [2]. Imaging concentrators, formed for instance by a single lens, compete advantageously when the desired concentration is small because of the availability of its elements.

The most well known nonimaging concentrator is the compound parabolic concentrator (CPC). The denominations CPC uses to group a large variety of concentrators designed with the method in [2]. Among these concentrators there is one, also called CPC, which is used to represent the set. The cross-section of this concentrator is formed by two mirrored parabolic branches. The total transmission of the CPC of revolution is above 95% for the whole range of acceptance angles (for $\theta < 5$ deg, the total transmission is about 95%). Its most important disadvantage is its thickness for small acceptance angles. The ratio of thickness to entry-aperture diameter of a CPC with a small acceptance angle is approximately $0.5(C_g)^{1/2}$, where C_g is given by (3.7); for instance, this ratio is 43 for a CPC of $n = 1.5$ and $\theta = 1$ deg, compared with the ratio of 0.28 of an XR or an RXI. When the acceptance angle is increased, the CPC increases its total transmission and strongly decreases its size, while the XR and the RXI decrease their total transmission and increase slightly their size; so, the CPC becomes a better candidate for large acceptance angles (above 10–15 deg).

There are other concentrators of the CPC family with a smaller size. For instance, the dielectric-filled CPC with curved entry aperture (DTIRC [17]) with a thickness to entry aperture ratio between 1 and 2 for acceptance angles above 10 deg. A concentrator of this type has recently been applied to an LED to form a high-efficiency lamp [18,19].

When the acceptance angle is smaller, the thickness of the concentrator can be reduced by combining an image-forming device like a parabolic mirror with a concentrator of the CPC family (usually a compound elliptical concentrator, CEC [2]). The highest levels of solar concentration have been achieved with combinations of parabolic mirrors and nonimaging concentrators, similar to the one mentioned previously (except that the nonimaging concentrator is not a CEC) [20]. Such combinations are thicker than the equivalent XR or RXI concentrators and their optical performance may be poorer. The theoretical limit of concentration can be attained with these two-stage systems if the imaging stage (parabolic mirror) is without aberrations (i.e., when the mirror forms the image of the source at the CEC entry aperture). This requires f numbers much greater than 0.287. For instance, consider a combination of an $f/2$ parabolic mirror with a dielectric-filled CEC ($n = 1.483$) to form a concentrator with design acceptance angle $\theta_a = 1$ deg. The geometrical concentration of the first stage is $C_{g1} = \sin^2\phi \cos^2\phi/\sin^2\theta_a = 181.76$ (see, for instance, [21]) where ϕ is the rim angle of the mirror, that is, $\phi = \tan^{-1}(1/2f)$.

The second stage concentration is approximately $C_{g2} = n^2/\sin^2\phi = 37.39$, so $C_{g3D} = C_{g1}C_{g2} = 6796$, which is 94% of the maximum concentration ($C_{g3D} \leq n^2/\sin^2\theta_a$). Considering this loss of concentration as a loss of total transmission and assuming that the total transmission of the CEC is similar to that of a CPC with acceptance angle $\theta_a \simeq \phi$ (i.e., $\mathbf{T}_{CEC} \simeq 0.962$ [2]), then the total transmission of the mirror-CEC combination is approximately 90% (i.e., 4.6% less than the nonimaging lens-mirror combination of Table 3.2). As the f number increases, the transmission-angle curve of the mirror-CEC combination gets more sharp, but the concentration of the first stage decreases, and then the concentration (and the size) of the second stage increases. This reduces the total transmission of the CEC. When f is very high, the shadow of the second stage on the mirror can be the cause of the decrease of the total transmission.

The rays suffer a single metallic reflection in both systems (assuming that the CEC can work with total internal reflection). The Fresnel reflections at the CEC entry aperture are approximately those corresponding to a beam impinging on its entry aperture at angles (with the normal to the entry aperture) below ϕ ($\phi \simeq$ 14.04 deg). The angular distribution at the lens entry aperture of the nonimaging lens-mirror combination is less homogeneous. At the center of the lens the beam is impinging on the lens at angles below 18.52 deg. Because of the quasi-spherical shape of the lens, and because the center of the lens is the point of the surface closest to the receiver center, it can be assumed that the rays will not form angles much greater than this with the normal to the lens surface; and so, concluding that, there will not be much difference in the Fresnel losses of both systems.

Optical losses depend on the type of materials used. Reflectivities between 85%–90% for the mirror reflection are easily achieved with an evaporated aluminum coating. Fresnel losses can be substantially reduced with an antireflection coating (this coating used to be too expensive for common applications when compared with the cost of the nonimaging concentrator). Absorption and dispersion losses not only depend on the type of dielectric material to be used, but also on the average ray length of the rays and, thus, it depends on the global size of the concentrator. Values of concentrator optical efficiencies between 70–90% can be expected using very common materials (for instance, poly-methyl-metacrilate as dielectric and Al coating as reflector) for average ray lengths not greater than 60 mm and acceptance angles greater than 1 deg. This means, for instance, that if there is an LED at the concentrator receiver position, then the power exiting the concentrator is 70–90% of the one exiting the LED. Another source of losses is the imperfection of the optical surface. These types of losses depend on the technology used to make the concentrator and are more important for small acceptance angles. Since the surfaces of nonimaging concentrators are aspheric, polymer injection molding used to be the most appropriate technology. In this case, imperfection of the optical surfaces depends strongly on the molding cycle time and on the maximum material thickness. At this time there are no measurements of the performance of XRc or RXIc because these concentrators have been designed very recently.

Another important feature of the XRc and RXIc is that there are no optical surfaces in contact with the receiver, as happens in the CPC (and CEC) designs whose mirror rims touch the receiver border. This feature should simplify the assembling of the concentrator and the receiver when this is small, for instance, when the device at the receiver is a photodiode or an LED and has to be glued to the dielectric material. An excess of glue can deteriorate part of the TIR reflector of the CEC because this reflector is in contact with the receiver border. Moreover, this part of the reflector subtends an important amount of etendue with the receiver, which is equivalent to saying that bad reflectivity in that region causes an important amount of losses. If there is not enough glue, bubbles of air can appear on the receiver surface, which would reflect by total internal reflection the radiation going to (or coming from) the receiver.

3.10 SUMMARY

A detailed description of the last developed method of design of nonimaging concentrators has been given. Nonimaging concentrators are optical systems designed to transfer optimally incoherent radiation from a source to a receiver and in such applications their performance is much better than the one obtained with imaging systems, and the number of optical elements needed uses to be fewer. One of these applications is the wireless transmission of IR radiation. The optical system used at the emitter generally collimates the radiation and the optical system at the receptor concentrates it.

The concentrators designed with the method described in this chapter have very interesting properties when their concentration is high. In particular, a very small concentrator thickness (ratio of concentrator thickness to aperture diameter around 0.28) and a performance close to the thermodynamic limit (total transmission above 95%) can theoretically be achieved. These characteristics are very interesting for wireless IR applications, which usually require a high concentration factor with the greatest possible acceptance angle. For medium to high acceptance angles (above 10 deg), the performance of the CPC family concentrators is better than the that of the XR and RXI, and also better than imaging concentrators.

REFERENCES

[1] Born, M., E. Wolf. *Principles of Optics*. Pergamon, 1964
[2] Welford, W. T., and R. Winston. *High Collection Nonimaging Optics*. San Diego: Academic, 1989.
[3] Smestad, G., H. Ries, R. Winston, and E. Yablonovitch. "The thermodynamic limits of light concentrators," *J. Sol. Energy Mater.*, Vol. 21, 1990, pp. 99–111.
[4] Ries, H., G. Smestad, and R. Winston. *Thermodynamics of light concentrators. Nonimaging Optics: Maximum Efficiency Light Transfer*. Edited by Roland Winston and Robert L. Holman. *Proc. SPIE 1528*, 1991, pp. 7–14.
[5] Miñano, J. C. "Design of Three-Dimensional Concentrators." In *Solar Cells and Optics for Photovoltaic Concentration*, Chapter 12, Edited by A. Luque. Bristol: Adam Hilger, 1989.
[6] Miñano, J. C. "Design of three-dimensional nonimaging concentrators with inhomogeneous media," *J. Opt. Soc. Am. A 3*, 1986, pp. 1360–1353.
[7] Miñano, J. C. "Two dimensional nonimaging concentrators with inhomogeneous media: a new look," *J. Opt. Soc. Am. A 2*, 1985, pp. 1826–1831.
[8] Bassett, I. M., W. T. Welford, and R. Winston. "Nonimaging optics for flux concentration," *Progress in Optics*, Vol. XXVII, 1989, pp. 161–226.
[9] Miñano, J. C. *Synthesis of concentrators in two-dimensional geometry*, in *Solar Cells and Optics for Photovoltaic Concentration*, Edited by A. Luque. Bristol: Adam Hilger, 1989, pp. 353–396.
[10] Winston, R., and W. T. Welford. "Geometrical vector flux and some new nonimaging concentrators," *J. Opt. Soc. Am.*, Vol. 69, 1979, pp. 532–536.
[11] Winston, R., and W. T. Welford. "Ideal flux concentrators as shapes that do not disturb the geometrical vector flux field: A new derivation of the compound parabolic concentrator," *J. Opt. Soc. Am.*, Vol. 69, 1979, pp. 536–539.

[12] Stavroudis, O. N. *The Optics of Rays, Wavefronts, and Caustics*, New York: Academic Press, 1972.

[13] Schulz, G. "Achromatic and sharp real imaging of a point by a single aspheric lens," *Applied Optics*, Vol. 22, 1983, pp. 3242–3248.

[14] Schulz, G. "Aspheric Surfaces." In *Progress in Optics*, Edited by E. Wolf, Vol. XXV, North Holland: Amsterdam, 1988, pp. 351–416.

[15] Welford, W. T., J. O'Gallagher, and R. Winston. "Axially symmetric nonimaging flux concentrators with the maximum theoretical concentration ratio," *J. Opt. Soc. Am. A 4*, 1987, pp. 66–68.

[16] Vidal, P. G., G. Almonacid, A. Luque, and J. C. Miñano. "Concentration Limits of primaries for cells immersed in optically dense media," *10th European Photovoltaic Solar Energy Conference*, Lisbon, 1991. Dordrecht: Kluwer, 1991, pp. 982–986.

[17] Ning, X., R. Winston, and J. O'Gallagher. "Dielectric totally internally reflecting concentrators," *Applied Optics*, Vol. 26, 1987, pp. 300.

[18] Gardner, R. C., G. E. Smith, and C. L. McLeod. *High efficiency lamp or light accepter*. US Patent 5,055,892. Oct., 1991.

[19] Gardner, R. C., D. E. Silvergate, G. P. Smestad, G. E. Smith, and J. F. Snyder. *Nonimaging light source*. US Patent 5,001,609. March 1991.

[20] Gleckman, P., J. O'Gallagher, and R. Winston. "Approaching the irradiance of the sun through nonimaging optics," *Optics News*, 1989, pp. 33–36.

[21] Winston, R. "Nonimaging Optics," *Sci. Am.*, Mar. 1991, pp. 76–81.

Chapter 4

Codification And Modulation Techniques For IR Wireless LANs

Angel Polo, Asunción Santamaría
E.T.S.I. Telecomunicación
Ciudad Universitaria s/n. 28040 Madrid, Spain

4.1 INTRODUCTION

Data transmission through optical wireless infrared (IR) links has an important field of application on indoor environments, such as local area networks (LANs). But the depth of penetration of electromagnetic waves is inversely proportional to the frequency, and IR signals cannot pass through most objects (such as walls). That is why IR wireless LANs (WLANs) use cellular architectures, where each cell is a room of a building [1] (see Fig. 4.1). Links connecting two or more cells can be wired. As stated in Chapter 1, three kinds of wireless links can be used within a single cell: point to point, diffuse, or quasi-diffuse.

This chapter shows the indoor environment transmission characteristics and the codification and modulation techniques to be used in these kind of systems. Section 4.2 shows a review of the standards on medium access methods for LANs. The possibility of transforming classically cabled LANs (CLANs) into WLANs is also presented. Section 4.3 presents an analysis of the most important problem in infrared communications: the noise due to ambient light. A study on optical power distribution in indoor environments is shown in Section 4.4. Modulation and codification techniques are discussed in Section 4.5 and finally, a summary is presented.

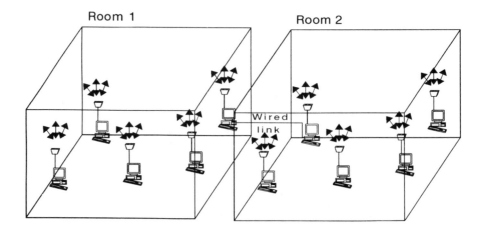

Figure 4.1 Cellular architecture for IR wireless LANs.

4.2 MEDIUM ACCESS METHODS FOR LANs

Conventional LANs abide by the IEEE rules, which include three medium access methods:

- IEEE Std. 802.3: Carrier Sense Multiple Access with Collision Detection (CSMA/CD) Access Method [2].
- IEEE Std. 802.4: Token Bus Access Method [3].
- IEEE Std. 802.5: Token Ring Access Method [4].

International organizations working on the standardization of WLANs have been presented in Chapter 1, so this section presents a brief explanation of classical methods and the study of their IR wireless implementation.

4.2.1 Carrier Sense Multiple Access with Collision Detection (CSMA/CD)

CSMA/CD local area networks, commercially know as Ethernet LANs, are the most common systems that can be found in office environments. Ethernet LANs have a bus structure with all the stations attached to the bus (cable). When a station wants to transmit, it listens to the cable, and if the cable is idle the transmission starts immediately. If the cable is busy, the station waits until it goes idle. If two or more stations simultaneously begin to transmit, a collision occurs. Then, stations stop their transmission and wait a random amount of time before repeating the whole process [5]. Data rates for Ethernet LANs are 1 and 10 Mbps with Manchester encoding [2].

Due to the bus architecture, wireless IR Ethernet LANs need diffuse or quasi-diffuse structures (point-to-point structures are not suitable for obvious reasons). However, available infrared-emitting diodes (IREDs) are able to warrant reliable IR-diffuse links up to 1 Mbps, that is, diffuse structures can be used for low-data-rate systems. So, quasi-diffuse structures are the only choice for 10 Mbps Ethernet WLANs [6].

4.2.2 Token Bus Multiple Access

Token bus networks have a bus topology where all stations are connected, listening to the medium every moment. The medium access method is a token-passing method, so logically the stations are organized into a ring. When the LAN is booted, a first task performs the logical ring initialization. After that, each station knows who its predecessor and following stations will be, and the highest numbered station may send the first frame; then it passes the token to its neighbor in the logical ring, as previously arranged.

Due to the bus topology that provides a broadcast medium, IR wireless token bus LANs can be implemented using diffuse and quasi-diffuse structures.

Data transmission, following the IEEE standard, is done by modulating the signal in a continuous-phase frequency shift-keying manner with carrier frequencies of 3.75 MHz and 6.25 MHz for a data rate of 1 Mbps.

Higher data rates (5 and 10 Mbps) use a coherent-phase frequency shift keying with carrier frequencies of 5 and 10 Mbps, or multilevel duobinary-amplitude modulate-phase shift keying.

Data rates oblige us to reject diffuse structures (as previously explained in describing CSMA/CD LANs), making quasi-diffuse structures the best option.

4.2.3 Token Ring Multiple Access

Token ring networks are a collection of individual point-to-point links that form a circle (ring) where each station is connected to its two neighbor stations, its preceding one and its following one.

In a token ring a special bit pattern, called the token, circulates around the ring. When a station wants to transmit, it removes the token from the ring before transmitting, delivering the token to the ring when the transmission is finished. There is only one token, so only one station can transmit in a given instant. Following this method, all stations will at some instant have their transmission right guaranteed.

Each station listens to the transmission sent from its preceding station and transmits to the following station in the ring, so diffuse and quasi-diffuse links are not a good choice for IR wireless token ring implementation. Point-to-point IR links between neighboring stations are the only method for establishing a wireless-ring topology. Precise alignments are required and the mobility property of wireless LANs gets lost. Additional problems appear when a new station wants to belong to the ring because link realignments must be done. Obstacles in the line of sight of receivers might be scrupulously avoided.

4.3 NOISE AND INTERFERENCE DUE TO BACKGROUND LIGHT IN WIRELESS IR TRANSMISSION

In wireless IR links the photodiode is exposed to the ambient light, introducing additional noise and interference to the input circuit of the receiver amplifier, particularly in diffuse and quasi-diffuse links where the incoming radiation is received from a wide field of view (FOV).

The most common ambient light sources are daylight, tungsten, and fluorescent lamps. Their relative power spectral densities are shown in Figure 4.2. In general there is a high level of stationary ambient light generating a dc photocurrent and white shot noise in the photodiode. In addition, artificial light sources also emit rapidly fluctuating components associated with the higher harmonics of the main frequency [7]. These frequency components generate a colored noise in the photodiode and, in the case of fluorescent lamps, create an interference problem.

4.3.1 Noise Due to Ambient Light

As stated previously, two types of noise are generated in the photodiode: a white shot noise plus $1/f$ noise generated by the dc photocurrent; and a colored noise generated by background light fluctuations and associated harmonics.

The spectral density of the shot noise current is given by the Schottky formula [8]:

$$\langle i_{sh}^2 \rangle = 2qI_b(1 + f_L/f)(A^2/Hz) \tag{4.1}$$

where $q = 1.6 \cdot 10^{-19}$ C is the electron charge, f_L is the empirical value of the break frequency where the two noise components are equal, and I_b is the dc background light photocurrent:

$$I_b = \frac{h \cdot \nu}{\eta \cdot q} \cdot P_b \tag{4.2}$$

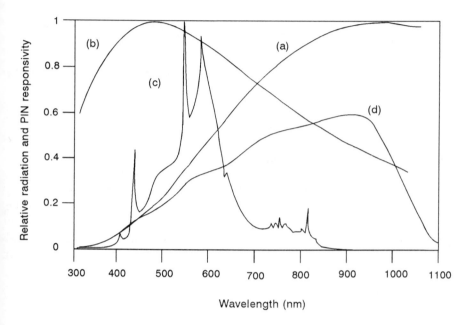

Figure 4.2 Relative energy spectral distribution and *p-i-n* responsivity: (a) tungsten lamp, (b) daylight, (c) fluorescent lamp, (d) *p-i-n* responsivity.

where P_b is the average optical power falling upon the detector from the ambient light source, η the photodiode quantum efficiency, v the optical frequency, and h Planck's constant.

The spectral density of the colored noise current has been found to be related to the electric spectrum of the ambient light (see following section). No theoretical expression of this noise has been available until now. The following relation agrees with the experiment in predicting the noise density:

$$\langle i_{cn}^2 \rangle = 2qA10^6 |I_b(f)| \ (A^2/Hz) \tag{4.3}$$

where $I_b(f)$ is the Fourier Transform of the photocurrent generated by ambient light fluctuations, and A is an experimentally derived constant found to be $2 < A < 3$.

Thus, the spectral density of the total current noise generated by ambient illumination is:

$$\langle i_{tn}^2 \rangle = \langle i_{sh}^2 \rangle + \langle i_{cn}^2 \rangle \tag{4.4}$$

The second term in the above equation dominates for frequencies where $H_p(f)$ is not negligible (up to 2 kHz for tungsten lamps, and up to 100 kHz for fluorescent lamps).

To compute the total noise at the receiver, bias network and amplifier noises have to be taken into account [9].

4.3.2 Interference Caused by Artificial Ambient Light Sources

Interference arising from artificial light sources, caused by harmonics of the supply frequency (50–60 Hz), produces a very severe depreciation in the detected signal. Figure 4.3 shows the spectrum of light fluctuations for tungsten and fluorescent lamps. The tungsten lamp has a power of 100W and it is placed at 1m from the photodiode (PD). As depicted, the interference is stronger for the fluorescent lamps. Therefore, in the presence of fluorescent light, optical filtering, electrical filtering, modulation techniques, or a combination of these have to be used in order to minimize the effects of noise and interference.

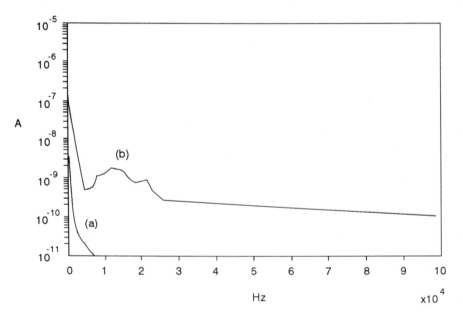

Figure 4.3 Ambient light electric spectrum: (a) standard tungsten lamp; (b) office with fluorescent lamps (1,000 Lux illumination level).

4.3.3 Noise and Interference Reduction

4.3.3.1 Optical Filtering

Part of the incident ambient light reaching the photodetector may be blocked by an optical filter. Two types of filters may be used: optical interference filters with a pass band corresponding to the LED source bandwidth and absorption filters blocking only the visible part of the spectrum. From the optical point of view, the best results are obtained from interference filters. In the case of diffuse and quasi-diffuse links however, its suitability is low for several reasons: first, the maximum transmission in the bandpass area is lower than 40%; second, its optical response changes when the angle of incidence of light is not 90 deg; third, they are very fragile components.

Thus, absorption filters have to be used. An effective and inexpensive filter is obtained by developing a slidefilm frame to a black target. Figure 4.4 shows this type of filter's response. With this filter, a 13-dB reduction is obtained in the optical power reaching the photodiode from fluorescent lamps, 3 dB for the tungsten lamp, and 5 dB for daylight. The attenuation at the LED wavelength is only 1.3 dB.

Other authors [8, 10] have proposed an absorption filter obtained by developing an unexposed color film. Although this filter offers a substantial reduction of ambient light, the attenuation introduced at the LED wavelength is higher than for the slide-film filter.

4.3.3.2 Electrical Filtering

As discussed in this section, the colored noise and interference components are located in the low-frequency band. Therefore, they can be filtered out by the use of a high-pass electrical filter.

Some authors have proposed electrical filters [8, 10, 11] for wireless IR applications. All of these filters are suitable for high-impedance amplifier configurations. Because transimpedance configuration has to be used for diffuse and quasi-diffuse systems in order to obtain a wider dynamic range, other types of high-pass filters have to be designed.

4.3.3.3 Modulation Techniques

Noise and interference from fluorescent light dominates receiver performance in the frequency region below 100 kHz. For this reason, baseband transmission is not considered adequate to ensure stable signal transmission and subcarrier waves with

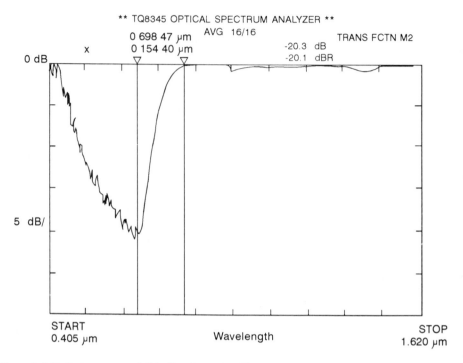

Figure 4.4 Optical response of slide-film absorption filter.

a frequency beyond a few-hundred Hertz should be used. The use of a modulated carrier allows the signal spectrum to be shifted away from the spectrum of ambient light fluctuations. Possible applications of different modulation techniques are analyzed in Section 4.5.

4.4 SIGNAL LEVEL ON IR WIRELESS LINKS

One of the more important parameters that system designers must keep in mind is the signal-to-noise ratio (SNR) at the system receiver . In Section 4.3 the noise power in wireless IR indoor channels has been explained. This section presents the study of signal power, focusing on quasi-diffuse links [12]. Studies about point-to-point links will be presented in Chapter 5. Studies about diffuse structures are omitted because, due to the limitations of the IREDs, they are inadequate for high-data-rate applications (like LANs).

IR quasi-diffuse WLANs work with a passive reflector that is usually placed at the center of the ceiling of the room. All of the equipment connected to the

WLAN points toward the reflector, which has a Lambertian reflection pattern. Optical signals transmitted are broadcasted and receivers in the covered area receive the signal after one reflection (optical signal contributions received with more than one reflection are negligible). Optical transmitters are composed of an IRED and an optical concentrator (lens) that allows directed beams to be obtained. Optical receivers are composed of a photodetector and an optical concentrator to increase the optical power received.

Optical losses in these systems can be classified into losses at the transmitter side (through the path) and at the receiver side. Losses at the transmitter side comprise (see Fig. 4.5):

- *Beam opening loss (BOL).* Due to the IRED angular emission profile, part of the transmitted power is lost. This is the "beam opening loss," which can be calculated as: $BOL = 10 \cdot \log(P_{LED}/P_{IL})$
- *Lens losses (LL).* Using lenses, not all the power that impinges on the lens is transmitted. A part is lost due to scattering and absorption effects. LL can be calculated as: $LL = 10 \cdot \log(P_{IL}/P_{TL})$

Losses through the path comprise (see Fig. 4.6):

- *Propagation losses (PL).* Due to atmospheric attenuation. PL can be calculated as: $PL = 10 \cdot \log(P_{TL}/P_{IR})$ but quasi-diffuse systems are intended to

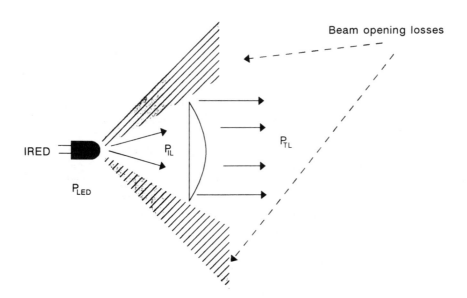

Figure 4.5 Losses at the transmitter side.

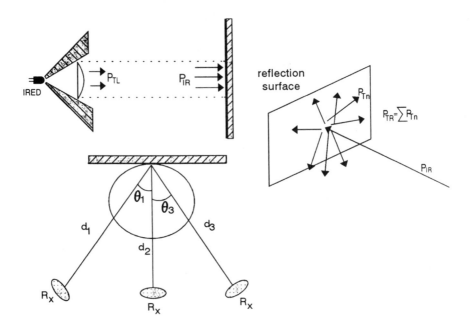

Figure 4.6 Losses through the path.

be used in an indoor environment. Atmospheric conditions in this environment yield a negligible propagation loss in comparison with other losses present at the system. Therefore, a good approach is to consider that $PL = 0$ dB.

- *Reflection losses (RL).* A part of the power that reaches the reflection surface is absorbed. This absorption is computed as power loss, and can be calculated as: $RL = 10 \cdot \log(P_{IR}/P_{TR})$
- *Placement losses.* Optical power received by a given set of equipment depends on two factors, the distance from the reflector (d) and the angle with the perpendicular to the reflection surface (θ) (figure 4.6). Due to the reflector Lambertian radiation intensity pattern, all the receivers placed at the circumference of θ angle and distance d will receive the same optical power. Then, placement losses are divided into:
  - *linear displacement loss (LDL),* calculated as:
    $LDL = 10 \cdot \log(P_{TR}/P_{Rx}(d))_{\theta = cte}$
  - *angular displacement loss (ADL),* calculated as:
    $ADL = 10 \cdot \log(P_{Rx}(\theta = 0)/P_{Rx}(\theta))_{d = cte}$

At the receiver, side the gain due to the use of optical concentrators must be computed. This *concentrator's gain (CG)* can be calculated as:

$$CG = 10 \log(\text{concentrator's surface/photodiode's surface})$$

From the previous statement we can conclude in the next expression for the optical power budget:

$$10 \cdot \log(P_{LED}/P_{Rx}) = BOL + LL + PL + RL + LDL + ADL \; CG$$

4.4.1 Example

In order to clarify the influence of each term of the power budget, it is presented the optical power budget in a quasidiffuse link at the following supposed environment: The room size is 10m by 10m by 3m, where all the IR equipment is placed 1m from the ceiling to avoid transmission errors due to obstacles on the transmitter line of sight. A small area at the center of a white-painted ceiling is used as the scatterer surface. All the equipment is pointed at this area and has these other features:

1. A transmitter lens, made in methacrylate with 7-cm diameter, 12-mm thickness, and 11-cm focal distance.
2. An IRED located at the lens focus, with a 12° halfpower angle.
3. A reflector, the ceiling of the room with a reflection coefficient $\rho = 0.72$.
4. A receiver concentrator like the transmitter lens.
5. A photodiode with a surface of 10 mm².

The best placement in this environment, from the point of view of the optical power received, is for the equipment located under the reflector ($d = 1m$ and $\theta = 0$). The worst case is for the equipment placed at a corner of the room ($d = 7.1m$ and $\theta = 81.87$).

The power budget for the best-case equipment is: $10 \cdot \log(PLED/PRx) = BOL + LL + PL + RL + LDL + ADL \; CG = 22.74$ dB, and for the worst-case equipment is: $10 \cdot \log(PLED/PRx) = BOL + LL + PL + RL + LDL + ADL \; CG = 48.25$ dB.

A three-dimensional map of the received power in the room, relative to 0 dB transmitted is shown in figure 4.7.

It should be emphasized that the most restrictive term in the power budget is *LDL* in both the best and worst cases. An *ADL* term becomes more important on increasing θ. For large angles (near the room corner) this is the second most

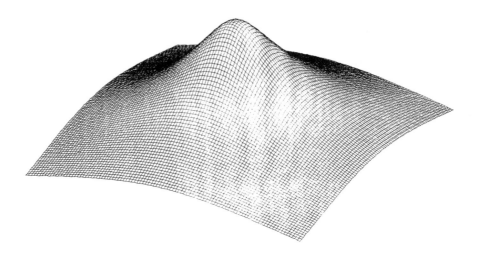

Figure 4.7 Three-dimensional map of received power.

important term. *PL* and *LL* are negligible terms, and an important improvement can be achieved by using optical concentrators. It is important to note the difference between the best and worst case (26 dB). Quasi-diffuse systems using the same hardware independent of the equipment placement must be able to work with a wide dynamic range. This is the most stringent requirement for the design.

4.5 MODULATION AND CODIFICATION TECHNIQUES FOR IR WLANs

This section presents a study about modulation and codification techniques that can be used in IR wireless links. On-off keying, frequency shift keying, and Multisubcarrier modulation will be analyzed together with Manchester, Miller, and NRZ codification methods.

4.5.1 On-Off Keying

On-off keying (OOK) is the simplest modulation technique for indoor IR wireless links. The intensity of the transmitted lightwave is directly modulated by data to be transmitted in such a way that symbol "1" turns the emitter on (state ON) and symbol "0" turns it off (state OFF). The most important advantage of this technique is the simplicity of the transmitter and receiver design (see Fig. 4.8).

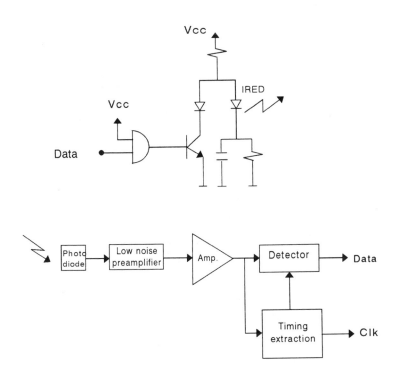

Figure 4.8 OOK transmitter and receiver circuits.

Coding methods must be used together with OOK modulation because it is necessary to discern between the transmitter state OFF when a symbol "0" has been sent and the transmitter state IDLE when it is inactive.

Coding methods presented here are Manchester code, Miller code and nonreturn zero (NRZ) with 4B/5B code.

4.5.1.1 Manchester Code

Symbol "1" is encoded as a falling edge in the center of the symbol time (T_b), and symbol "0" is encoded as a rising edge in the center of the symbol time (T_b). This coding method assures that a transition is present in each symbol (see Fig. 4.9). Therefore this code allows an easy timing extraction, but the necessary transmission bandwidth is twice the quantity required for NRZ code.

Using Manchester-encoded signals a transition occurs each T_b sec at most, and each $T_b/2$ sec as a minimum.

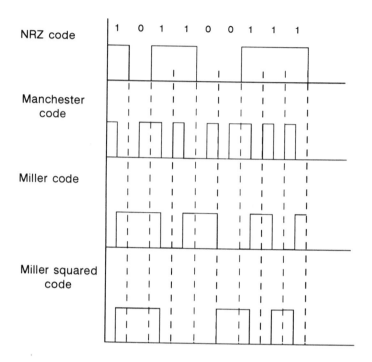

Figure 4.9. Coding methods.

The spectrum of the Manchester-encoded signal includes weak low-frequency components and it is possible to filter out interferences at these frequencies using a high-pass filter.

4.5.1.2 Miller Code

Symbol "1" is encoded as a transition in the center of the symbol time (T_b) and symbol "0" is encoded as no transition. But, in the case where a "0" is followed by another "0" a transition is present in the first "0" (see Fig. 4.9). Using Miller-encoded signals, a transition occurs each $2 \cdot T_b$ sec at most, and each T_b sec as a minimum.

Squared Miller code uses the same rule as Miller code, but when there is an isolated symbol "0" followed by an odd number of symbols "1", the transition on the last "1" is removed. In this case, a transition is present on the signal each $3 \cdot T_b$

sec at most, and each T_b sec as a minimum. Therefore, Miller coding provides a signal with less transition than Manchester-encoded signals (see Fig. 4.9).

4.5.1.3 NRZ Code

Symbol "1" is represented as a "high" level during all the bit time and symbol "0" as a "low" level during all the bit time. Redundant bits are necessary to assure the presence of transitions in the transmitted signal that allow timing extraction at the receiver. A good choice is to use NRZ code with 4B/5B codification because all the frame structures of standard LANs are based on 8-bit groups (bytes) and it is easy to divide an 8-bit group into two 4bit length halves. The number of transitions in a NRZ + 4B/5B coded signal is fewer than in Manchester or Miller codes, and redundancy bits assure a good timing extraction. Using 4B/5B codification obliges a link data rate equal to (source data rate)*5/4.

The codification method must be chosen according to data rate and transmission bandwidth available in each application.

Point-to-point and quasi-diffuse links do not present additional complexity. Data rate is limited by the IRED's switching times. But diffuse links present a power penalty due to intersymbol interference (ISI), which gets more important as the data rate is increased. Therefore, decision feedback equalizers (DFE) are recommended at the diffuse link receivers [13].

4.5.2 Multisubcarrier Modulation

J. R. Barry et al. have proposed multisubcarrier modulation for IR wireless transmission [13].

Using this modulation scheme, the available bandwidth is divided into a number of sub-bands. One link is established across each sub-band in such a way that data rate at these links is reduced.

Suppose that the source bit rate is R_s bps. Bits from the source are grouped into blocks of L bits length that will be transmitted with a block rate R_b (blocks/s); so, $R_s = L \cdot R_b$ bps (see Fig. 4.10).

Each L bits block is divided into M interleaved bit streams, and each stream modulates one carrier. All the modulated carriers are summed, and the result of this sum modulates the intensity of the transmitted lightwave.

With this method, the total bandwidth is divided into M sub-bands, and each substream uses only $1/M$ of the total bandwidth. Therefore, multisubcarrier systems are less susceptible to ISI.

Modulated carriers must be separated in the receiver before demodulation. This can be done using frequency division multiplexing technology [14]. A set of

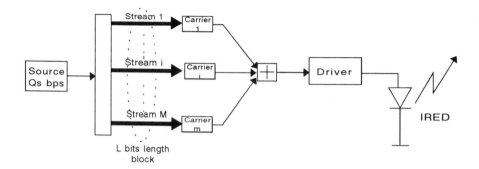

Figure 4.10 Multisubcarrier transmitter.

filters can be used to separate the bands at the receiver. To make the implementation of this kind of filter easy, each signal must use a bandwidth

$$B_n = f_N(1 + \alpha) \tag{4.5}$$

where f_N is the Nyquist frequency, obtaining a band usage efficiency of

$$\frac{f_N}{\Delta f_N} = \frac{1}{(1 + \alpha)} \tag{4.6}$$

The number of subcarriers M must be chosen maintaining an ISI penalty that negligible in each subsystem. The number of bits per subcarrier and the power per subcarrier can be adjusted to the SNR expected at each carrier frequency.

Advantages of multisubcarrier modulation systems are the bandwidth efficiency, the capacity to counter multipath effects, and the fact that low-frequency harmonics of fluorescent light are avoided due to the signal spectrum that is moved away from dc.

4.5.3 Frequency Shift Keying (FSK)

In a FSK system, two sinusoidal waves of the same amplitude Ac, but different frequencies f_1 and f_2, are used to represent symbols "0" and "1" respectively. The binary FSK wave can be expressed as:

$$s(t) = \begin{cases} A_c \cos 2\pi f_1 t & \text{for symbol "0"} \\ A_c \cos 2\pi f_2 t & \text{for symbol "1"} \end{cases} \tag{4.7}$$

To generate a FSK wave, the incoming data are applied to a frequency modulator. The modulator changes the transmitted frequency when the input changes from one voltage level to another. The FSK-modulated signal modulates the intensity of the lightwave to be transmitted (See Fig. 4.11).

It must be noted that f_1 and f_2 must be chosen larger than the bit rate, $1/T_b$, where T_b is the bit duration.

Two kind of demodulators can be used for the demodulation of FSK waves: coherent FSK demodulators and noncoherent FSK demodulators.

4.5.3.1 Coherent FSK Demodulator

Coherent FSK demodulators (Fig. 4.12) consist of two correlators that are individually tuned to the two different carrier frequencies f_1 and f_2. The decision device

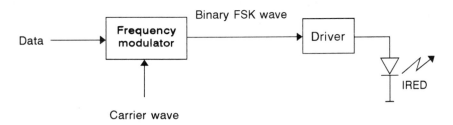

Figure 4.11 FSK modulator.

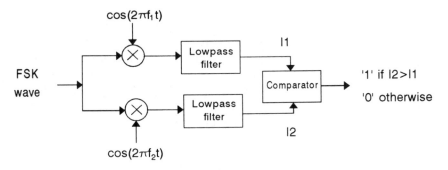

Figure 4.12 Coherent FSK demodulator.

compares the two correlator outputs and if the output l_2 produced in the lower path is greater than the output l_1 produced in the upper path the detector makes a decision in favor of symbol "1." Otherwise, it decides in favor of symbol "0."

Coherent demodulators need two forms of synchronization for their operation, one that ensures that the carrier generated at the receiver is locked in phase with the carrier used at the transmitter (phase synchronization), and the second that ensures the proper timing of the decision operation at the receiver side with respect to the switching times in the input data stream to the modulator at the transmitter side (timing synchronization).

The probability of error for coherent FSK detectors is given by [15]

$$P_e = \frac{1}{2} \, erfc \left(\sqrt{\frac{E_b(1 - \rho)}{2N_0}} \right) \tag{4.8}$$

when the channel noise is white and Gaussian, with zero mean and power spectral density $N_0/2$. E_b is the signal energy per bit,

$$
\begin{aligned}
E_b &= \int_0^{T_b} A_c^2 \cos^2(2\pi f_1) dt \\
&= \int_0^{T_b} A_c^2 \cos^2(2\pi f_2) dt = \frac{A_c^2 T_b}{2}
\end{aligned}
\tag{4.9}
$$

and it is assumed that the symbols "0" and "1" are sent with the same probability, that is, 1/2.

4.5.3.2 Noncoherent FSK Demodulator

Noncoherent FSK demodulators (Fig. 4.13) consist of two matched filters followed by two envelope detectors. One filter is matched to $A_c \cdot \cos(2\pi f_1 t)$ (symbol "0"), and the other is matched to $A_c \cdot \cos(2\pi f_2 t)$ (symbol "1"). To remove the dependence of the matched filter output on the phase of the received signal, envelope detectors are used. Outputs of these envelope detectors are sampled every T_b seconds. Calling these samples as l_1 and l_2, if $l_2 > l_1$ the receiver chooses symbol "1." Otherwise, it chooses symbol "0." The average probability of error in these receivers is [15]:

$$P_e = \frac{1}{2} \exp\left(-\frac{E_b}{2N_0} \right) \tag{4.10}$$

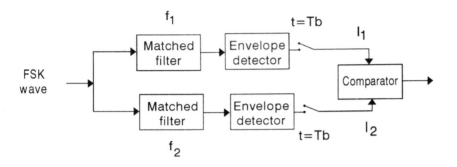

Figure 4.13 Noncoherent FSK demodulator.

where it is assumed that the channel noise is white and Gaussian, with zero mean and power-spectral density $N_0/2$ and E_b being the signal energy per bit.

4.6 SUMMARY

A review of IR wireless links has been presented and, focusing on LANs, a study of the access methods that can be implemented using IR wireless transmission has been done. One of the most important obstacles, the noise due to the ambient light, has been characterized. A study about optical power distribution in quasi-diffuse links has been presented. We have also presented the modulation and codification techniques that can improve the features of the received signal, keeping in mind the limitation from the commercially available optoelectronic devices.

At present, there is much work to be done before these systems become practical. There is an important limitation on the optical power available on commercial IREDs, and it is necessary to the development of low-noise amplifiers at the receiver side. More analysis is also required to determine the best modulation scheme.

Current trends indicate a decrease in hardware costs for optoelectronic devices and associated circuits, whereas costs for copper cables and installation and maintenance of wired systems are not showing a reduction. There are new applications of IR wireless local area networks that can offer several advantages.

REFERENCES

[1] Pahlavan, K. "Wireless Intraoffice Networks," *ACM Trans. on Office Information Systems*, Vol. 6, No. 3, July 1988, pp. 277–302.

[2] ANSI/IEEE Std. 802.3, *Part 3: Carrier Sense Multiple Access with Collision Detection (CSMA/CD) access method and physical Layer Specifications*, The IEEE, Inc., February 24, 1989. ISBN: 155937-005-X.

[3] ANSI/IEEE Std. 802.4, *Token-Passing Bus Access Method and Physical Layer Specifications*, The IEEE, Inc., February 25, 1985. ISBN: 0-471-82750-9.

[4] ANSI/IEEE Std. 802.5, *Token-Ring Access Method and Physical Layer Specifications*, The IEEE, Inc., April 29, 1985. ISBN:0-471-82996-X.

[5] Tanenbaum, A. *Computer Networks*, 2d ed., London: Prentice Hall, 1988.

[6] Santamaría, A., J. L. Muñoz, M. J. Betancor, F. J. Gabiola, and F. J. López-Hernández, "Optical-Electrical Interface for IR Wireless Ethernet Local Area Network," *Proceedings of the 16th Conference on Local Computer Networks*, IEEE Computer Society, Minneapolis, Minnesota, October 14–17, 1991, pp. 273–275.

[7] Gfeller, F. R., and U. Bapst. "Wireless In-House Data Communication via Diffuse Infrared Radiation," *IRE Proc.*, Vol. 67, No. 11, November 1979, pp. 1474–1486.

[8] Davenport, W., and W. Root. *An Introduction to the Theory of Random Signals and Noise*, New York: IEEE Press, 1987.

[9] Personick, S. D. "Receiver Design for Digital Fiber Optic Communication Systems," *Bell Syst. Tech. J.*, Vol. 52, 1973, pp. 843–874.

[10] Georgopoulos, C. J. "Filtering Techniques for Free Channel Infrared Detection in Closed Working Areas," *SPIE, Vol. 728: Optics, Illumination and Image Sensing for Machine Vision*, 1986, pp. 272–278.

[11] Roviras, D., M. Lescure, M. Duarie, and J. Boucher. "Suppresion of Environmental Interference in Infrared In-House Communications," *L'Onde Electrique*, Vol. 69, No. 1, January 1989, pp. 40–45.

[12] Santamaría, A., A. Polo, F. J. Gabiola, M. J. Betancor, and F. J. López Hernández. "Optical Power Budget on Quasi-Diffuse IR WLANs," *Proc. of the Third IEEE Int. Symp. Personal, Indoor and Mobile Radio Communications*, October 19–21, 1992, Boston, Massachusetts, pp. 159–163.

[13] Barry, J. R., J. M. Kahn, E. A. Lee, and D. G. Messerschmitt, "High-Speed Nondirective Optical Communication for Wireless Network," *IEEE Network Magazine*, Vol. 5, No. 6, November 1991, pp. 44–54.

[14] John A. C. Bingham, "Multicarrier Modulation for Data Transmission: An Idea Whose Time Has Come," *IEEE Communications Magazine*, May 1990, pp. 5–14.

[15] Haykin, S. *An Introduction to Analog and Digital Communications*, New York: John Wiley & Sons, Inc., 1989.

Chapter 5
Long-Haul Atmospheric Infrared Links

Pedro Menéndez-Valdés
E.T.S.I. Telecomunicación
Ciudad Universitaria s/n. 28040 Madrid, Spain

5.1 INTRODUCTION

Although IR wireless links are most commonly recommended for indoors communications, spanning long outdoor may sometimes be required for specific purposes. An example of this is a local network linking together a set of buildings of a given corporation (each of them containing a cabled subnetwork), distributed in an area with several kilometers in radius. Since atmospheric optical links are subject to adverse weather conditions that severely degrade the transmission quality, they are certainly disadvantageous compared to optical cable or radio distribution (including microwave links). However, some circumstances may make it advantageous to use IR-directive beams to link specific parts of the networks as non-permanent links, or for situations where cabling is not possible due to physical restrictions, problems related to the right of way, or local regulations of cabling.

An experiment of this kind was carried out in the Polytechnic University of Madrid, starting in 1983 [1]. At that time, a part of the data network of the University at the "Ciudad Universitaria" campus northwest of Madrid was implemented as a star network of point-to-point IR open-air links—up to 2.5-km long, with the distribution center being the Civil Engineering School. The emitters were GaAlAs laser diodes, and simple refractive optics were used both for collimation in the emitter side and for collection in the receiver [2]. Experimental atmospheric optical-ring networks have also been developed recently in other similar environments [3].

Optical technologies offer well-known advantages over microwaves for data transmission systems. The most outstanding advantages of these systems in open-air applications are the higher data rates achievable, the improvement in privacy due to the higher directivity of light, even for smaller antenna sizes, and the use of a spectral region where there are no reserved bands. The reasons why atmospheric optical LANs (AOLANs) are not commonly used are the impossibility to provide all-weather operation, and users not having a knowledge of established models to determine the network performance. Since the author can do nothing to modify meteorology, his contribution will be to try to define a comprehensive link performance model useful to AOLAN designers. The use of medium or long atmospheric spans as the communication channel introduces a set of random, weatherdependent disturbances that affect the quality of the communication. This chapter will concentrate on discussing those effects and provide information to evaluate link budgets of atmospheric links. On the other hand, the architectures and protocols of the AOLANs are not specific for the atmospheric channel, and the optical antennas, the electronic drivers of the light source (laser or LED), and the front-end receivers are not different from those described in other parts of this book. Therefore, none of these topics will be presented here.

5.2 THE ATMOSPHERIC CHANNEL

For usual communication purposes, where the involved beam intensity is not high enough to induce nonlinear effects, there are two phenomena that make the atmospheric channel different from free space: refraction and attenuation. Although the refractive index of air is very close to unity, relative variations in long paths (mostly due to air density decreasing with height) can lead to small angular variations in the beam direction, which may be comparable with the divergence of the beam. On the other hand, absorption lines of atmospheric gases (mostly water vapor) and scattering by gas molecules, and more noticeably by aerosols, introduce an additional source of losses to be taken into account in the link budget. However, the main characteristic of the atmospheric channel is its randomness. Both phenomena can have random nature and this fact must be accounted for in the link failure analysis or the communication error rate.

Random variations of attenuation are associated with climatic variations, therefore they are very slow and usually considered in terms of the percentage of time in which the link can be established: The atmosphere is not an all-weather channel in the optical (IR or visible) band.

However, in "good weather" conditions there are unavoidable random variations of the local value of the refractive index with impact on the quality of the communications link. Turbulent flux of hot air is usually induced near the ground due to its heating during the day. The ascending flux of heated air is characterized by small-sized eddies where temperature gradients are large enough to induce sensitive random-refraction effects. The farther from the ground, the larger the eddies and the lower the temperature gradients, and therefore the effect of turbulences on random refraction decreases. Turbulence is also intensive in the neighborhood of thermal inversions, where two atmospheric layers with different temperatures slide one over the other, producing a turbulent interface.

Using a wave-optics approach, turbulences induce wavefront distortion. From a ray-optics point of view, rays of light follow randomly varying ways, and constructive or destructive interferences may occur in ray crossings. The effect should be only noticeable if actual wave fronts are present (i.e., if there is some spatial coherence of light). This happens at any relevant outdoors-link distance where the light source (whatever its nature and for reasonable antenna sizes) is perceived at the receiver side as a point source, and coherent spherical waves would be observed unless some channel distortions are present.

In this chapter we will discuss the basics of all of the relevant atmospheric perturbations and their effect on IR beams with the exception of deterministic refraction, which may be considered negligible for the path lengths relevant to LAN applications. Comprehensive and detailed analysis of the many atmospheric perturbations can be found in specialized books [4,5].

5.3 ATMOSPHERIC ATTENUATION

Radiation-matter interactions in the channel at optical frequencies are unavoidably important due to the high energy transported by optical photons (in comparison to radio photons). Some atmospheric gas molecules absorb IR energy in exchange for modifying its internal vibration state. At the same time, the atmospheric molecules act as scattering centers to the incoming radiation. Finally, the atmosphere is a turbid fluid. Particles in suspension (aerosols) degrade the atmospheric transparency by acting as absorbing pollutants and, more important, by producing a strong scattering of the incoming radiation. The total attenuation coefficient (per length unit) is therefore the sum of the contributions of molecular absorption α_m, molecular scattering α_R, and aerosol extinction α_a, the last one including the effects of both absorption and scattering by particles in suspension:

$$\alpha = \alpha_m + \alpha_R + \alpha_a \tag{5.1}$$

5.3.1 Gas Absorption

It is known from quantum mechanics that a molecule can change its vibration state, absorbing or emitting an infrared photon of the adequate energy (i.e., wavelength). Fortunately for light, the Earth's atmosphere is mostly composed by atomic gases (Ar), which does not form molecules, or diatomic molecules with identical atoms (N_2, O_2) where vibration transitions are not allowed.

The absorbing gases that can be found in the atmosphere in the largest concentration are water vapor and carbon dioxide. Since liquid water is a major component of the Earth, evaporation and condensation processes make the water content of the atmosphere very unstable, its concentration varying along with time and space. Precisely for this reason, water concentration is also a common meteorologic measurement and the amount of absorbing water-vapor molecules can be determined from absolute humidity and atmospheric pressure. The CO_2 also presents a variable concentration, being more important in urban areas and having a smooth variation from day to night in forests [6]. It is, however, admitted that far away from big cities its concentration is in the order of 330 ppm. Other absorbers (CO, O_3, CH_4, N_2O, NO, NO_2) are present in the atmosphere in very low concentrations [7,8].

Each absorption wavelength is a resonance around which an "absorption line" shapes the absorption spectrum. Each absorption line is defined by a "line strength" related to the probability of the transition and a "line width" related to the statistical uncertainties involved in the transition phenomenon, including the quantum uncertainty. Details on these issues may be found in any book on spectroscopy. Regarding the spectroscopy of atmospheric gases, *Atmospheric Radiation* [9] may be referenced, and *Laser Beams in the Atmosphere* [4] is a recommended reference regarding its emphasis on light propagation. With the exception of the high atmosphere where pressure is very low, the absorption line spectral broadening is dominated by pressure-induced molecule collisions, displaying a Lorentzian shape:

$$\alpha_m(v) = \frac{S}{\pi} \frac{\gamma}{(v - v_0)^2 + \gamma^2} \tag{5.2}$$

where S is the line intensity or line strength, v the wavenumber (i.e., the inverse of the wavelength), and γ is the line half-width, which depends on pressure and temperature. Usually, the linewidth γ_0 is measured at given pressure P_0 and temperature T_0, and then used to calculate its value at any other conditions P, T. For minor atmospheric constituents, this calculation can be simplified to

$$\gamma(P, T) = \gamma_0 \frac{P}{P_0} \sqrt{\frac{T_0}{T}} \tag{5.3}$$

The absorption lines of atmospheric absorbers are grouped in spectral bands related to vibration-rotation energy transitions. The atmosphere is completely opaque in some spectral regions where there is a high density of absorption lines. Some regions in between with a very low number of absorption lines are named "atmospheric windows," where optical transmission is possible. One of these atmospheric windows extends along all the visible and near IR where semiconductor lasers and LEDs may be used for communication purposes. The main absorber here is water vapor. Figure 5.1 shows the near-IR transmission at 1 km of clear atmosphere to optical sources with 1.5 nm of spectral width.

The absorption lines are extremely narrow (in the order of some pm), and they appear averaged in the graph over the emitter bandwidth. If a monochromatic laser diode is used, the spectrum must be resolved to a very high accuracy and care must be taken not to emit at exactly one absorption wavelength. As an example, Figure 5.2 shows, in a narrow spectral band, the atmospheric transmission at 10 km for a laser diode exhibiting a single Lorentzian emission line 1-MHz wide. Avoiding absorption lines is a little bit tricky since the emission wavelength of semiconductor lasers varies with temperature and current. Thermal and bias control is therefore necessary. To avoid such complexity, multiwavelength (FP) lasers are recommended for atmospheric links when the required power cannot be achieved with an LED.

The most complete atlas of atmospheric absorption lines is the Air Force Geophysics Laboratory (AFGL) Atmospheric Absorption Line Parameters Compilation [10,11], also known as HITRAN and updated yearly. AFGL at Hanscom

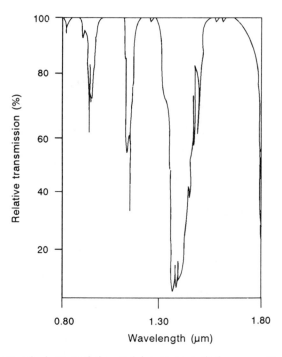

Figure 5.1 Sea-level atmospheric transmission at 1 km to an optical source with Gaussian optical spectrum 1.5-nm wide (FWHM).

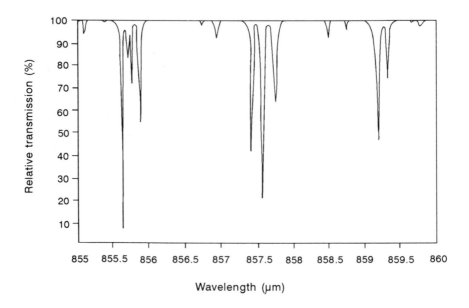

Figure 5.2 Sea-level atmospheric transmission at 10 km to an optical source with a Lorentzian spectrum 1-MHz wide (FWHM).

(Massachusetts) also developed weather-dependent models of low (20 cm^{-1}) and moderate (50 to 2 cm^{-1}) spectral resolution, which may be used to compute atmospheric transmission to broad-spectrum optical sources such as LEDs or multiwavelength semiconductor lasers [12]. Computer codes, which also include refraction and aerosol models, and radiance calculations for design of electrooptic detection systems are called LOWTRAN and ModTRAN, respectively. PC versions of both LOWTRAN and ModTRAN, and FASCODE (line-by-line spectral profiling of HITRAN for horizontal and slant paths), are commercially available from ONTAR Corporation and updated regularly.

5.3.2 Molecular Scattering

Electromagnetic waves propagating in material media induce polarization on the electric charge of the molecules. The resulting oscillating dipoles in turn behave as radiant antennas, scattering part of the incident energy all around. This phenomenon is known as Rayleigh scattering after John William Strutt, Baron of Rayleigh, who investigated the blue color of the sky. He showed that particles much smaller than the wavelength (such as gas molecules when compared to visible wavelengths) scatter part of the incident radiation with a scattering coefficient inversely proportional to the fourth power of the wavelength. If one accounts for all the scattering losses and the refractive-index dependence on atmospheric pressure and temperature, and on the wavelength, very accurate estimations of Rayleigh scattering in the near IR can be obtained by the formula [13]:

$$\alpha_R \approx 2.9154 \times 10^{-4} \frac{1 + 6.6 \times 10^{-3}/\lambda^2}{\lambda^4} \frac{P}{T} \tag{5.4}$$

with P in mb, T in K and λ in microns.

5.3.3 Aerosol Extinction

The field scattered from a plane wave impinging a spherical particle was first calculated by Mie by using a field expansion in spherical harmonics, over which the boundary conditions are easily applied [14]. One of the most interesting features of the work is that it is developed for the analysis of suspended metallic particles, which, from the optical point of view, correspond to materials with a large absorption. The approach to the problem of taking the metallic conductivity as an imaginary part of the dielectric permitivity makes it possible to compute the whole extinction as a single scattering problem involving a complex refractive index [15]. Thus, scattering, absorption, and total extinction can be computed simultaneously. Since, in real aerosols, particles may have different sizes, it is found useful to work with the scattering, absorption, and extinction efficiency factors, the last one being the sum of the other two and defined as

$$k_a = \frac{\alpha_a(r)}{\pi r^2} \tag{5.5}$$

where α_a is the aerosol extinction coefficient, and r the particle radius. The extinction efficiency factor can be expressed as a the sum of infinite terms, each computed from complex Bessel functions of the first and the second kind [16]. The convergence of the sums is very slow, however, and algorithms have been developed to compute the terms in a recursive manner, using inverse recursivity in some cases (i.e., estimating the number of significant terms, calculating the highest order one, and proceeding towards lower order terms in a recursive way) [17].

Figure 5.3 shows as an example the efficiency factor dependence on the particle size for the case of a spherical drop of water. The refractive index of water corresponds to its value measured at 4μm. Particles smaller than the wavelength scatter light at a rate which is dependent on the fourth power of the ratio r/λ, exactly as it was predicted by Rayleigh theory for gas molecules. For increasing

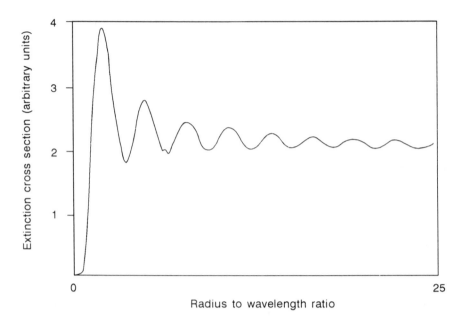

Figure 5.3 Mie extinction efficiency factor of light scattered by a spherical particle with complex refractive index 1.455–j0.005.

radius, the inverse dependence of extinction with the wavelength is still observed, although it smooths down below the quadratic dependence. This explains that fine-spread hazes present bluish scattering. The maximum scattering is observed when particle radii are in the order of the wavelength. In addition, the wavelength dependence of scattering shows very sharp variations in this region. Finally, for large particles, the extinction factor tends to a constant (wavelength independent) value. The white color of clouds and fogs, or finely divided powders such as flower, lime, or plaster, indicates that the particle size is significantly larger than the visible wavelengths.

All aerosols have particles of many different sizes, each of them showing its characteristic wavelength dependence of scattering. A particle-size distribution $f(r)$ represents the proportion of particles of each size. For a particle concentration of N particles per volume unit, the total attenuation of the aerosol is given by

$$\alpha_s(\lambda) = N \int_0^\infty \pi r^3 k_a(\lambda) f(r) dr \tag{5.6}$$

Aerosol composition depends very much on the geographic location, which determines the nature and amount of particles in suspension. Thus, rural aerosols are mostly constituted of particles of vegetal origin and dust in different concentrations, depending on the local flora and soil composition [18]. Urban aerosols have an important content of carbonaceous particles and other solid pollutants of industrial origin [19], and maritime aerosols not only depend on the continental mass transported to the location, but also on the water droplets generated by the wind on the sea surface [20]. Small-size ashes originated by volcanic eruptions or largemagnitude fires (for instance, those of oil wells burning during the Gulf War, 1992) can stay in the stratosphere as quasi-permanent aerosols for a considerable number of years [21]. In the same way, particles of continental origin may be transported long distances away from the particle source [22], making it difficult to predict the composition of local aerosols. Finally, fogs are originated by water condensation around hygroscopic particles, and therefore depend on the composition of the dry aerosol [16]. As an example, Figure 5.4 shows the extinction coefficient of coastal fogs and haze models at several atmospheric visibilities [23].

For visible light, the relation between aerosol extinction and atmospheric visibility expressed in km (vis) is

$$\alpha(\lambda = 0.55 \ \mu m) = \frac{3.912}{vis} \tag{5.7}$$

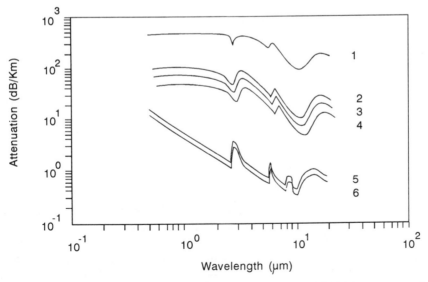

Figure 5.4 Attenuation coefficient versus wavelength for coastal fogs and thick hazes at meteorological visibilities of (1) 0.5 km, (2) 0.8 km, (3) 1 km, (4) 1.5 km, and light maritime hazes at relative humidities of (5) 90% and (6) 80%, according to the models of reference.

In the absence of dense fogs, atmospheric transmission in the near IR is usually slightly better than in the visible spectral region because of the wavelength dependence of extinction for hazes.

Precipitation is a particular kind of aerosol, characterized by a large particle size. Since particles are large when compared to any relevant infrared wavelength (i.e., any wavelength below 14 microns, where atmospheric windows exist), its contribution to the atmospheric attenuation is wavelength independent in all of the optical spectrum. Several models developed for different types of rainfall conclude that the types are proportional to a variable power of the precipitation rate R:

$$\alpha_a(\text{rainfall}) = A \times R^B \qquad (5.8)$$

where some values for the parameters A and B are expressed in Table 5.1. for R in mm/h and α_a in km^{-1}. The drop-size distributions of rain, however, depend on many local conditions that are difficult to predict. Therefore, the attenuation dependence on the precipitation rate has only an estimative value, and it is usually useful to find it empirically.

Table 5.1
Parameters of the Rainfall Attenuation Formula for Several Models of Raindrop Size Distribution

Rainfall Type	Parameter A	Parameter B	Reference
Continental rain	0.21	0.74	[4]
Rain (LOWTRAN model)	0.365	0.63	[24]
Drizzle	0.509	0.63	[25]
Widespread rain	0.319	0.63	[25]
Thunderstorm	0.163	0.63	[25]
Thunderstorm	0.401	0.63	[26]

5.4 TURBULENCE DISTORTION

Air density depends on the local temperature, and therefore air masses at different temperatures have different refractive indices. Hence, atmospheric turbulences induce random variations in the local index of refraction. A measure of these fluctuations is given by the index structure function, given by the mean square difference between the refractive indices of two points as a function of the distance between them:

$$D_n(\rho) = \langle [n(r_0 + \rho) - n(r_0)]^2 \rangle \qquad (5.9)$$

The Kolmogorov analysis of the turbulent flux predicts that the index structure function is proportional to the distance between points raised to two-thirds, the proportionality constant being called the structure parameter C_n^2 [27]:

$$D_n(\rho) = C_n^2 r^{2/3} \tag{5.10}$$

This means that the turbulence is isotropic (i.e., the variation of the refractive index between two points just depends on the distance between them, and not on their relative position on the wave front). The value of the structure parameter depends on the type of soil, the height of the path, and the local temperature. Typical values near the ground level are on the order of 10^{-15} m$^{-2/3}$ at 15°C and 10^{-14} m$^{-2/3}$ at 25°C.

Atmospheric turbulences distort the optical wave fronts. The measure of the wavefront distortion is the wave structure function, defined as the structure function of the natural logarithm of the field (i.e., the mean square variation of the logarithm of the field between two points as a function of the distance between them). For low turbulence, the Rytov theory applies [28,29], and the wave structure function is given by

$$D_w(\rho) = 2.914 k_0^2 \rho^{5/3} \int_0^L C_n^2(z) dz \tag{5.11}$$

for a plane wave, and

$$D_w(\rho) = 2.914 k_0^2 \rho^{5/3} \int_0^L C_n^2(z) \left(\frac{z}{L}\right)^{5/3} dz \tag{5.12}$$

for a spherical wave. Here ρ represents the distance between two points in a wave front at a distance L of the emitter, k is the wave number $2\pi/\lambda$, for λ the wavelength, and $C_n^2(z)$ is the distribution of the refractive index structure parameter along the beam path. Gaussian beams can be approximated by plane waves in the near field (i.e., close to the beam waist) and spherical waves in the far field.

The most direct effect of the wavefront distortion is the loss of coherence. Ideally, any coherent or incoherent optical source can be considered as a point source when observed from a long distance compared to its size. Therefore, far wave fronts are approximately spherical. In practice, due to the turbulence, the mutual coherence function is not a constant throughout spherical surfaces because of a nonconstant wavestructure function. Since the wave structure function is related to the logarithm of the field, it is easy to see that [28]

$$\Gamma(\rho) = \langle E(r + \rho) E^*(\rho) \rangle = E_0^2 e^{-D_w(\rho)/2} = E_0^2 e^{-(\rho/\rho_0)^{5/3}} \tag{5.13}$$

where

$$\rho_0 = \left(\frac{2}{D_w}\right)^{3/5} \rho \qquad (5.14)$$

which, according to (5.11) and (5.12), is a constant for a given path because the ρ-dependence is canceled. This magnitude is the coherence radius defined at $1/e$ coherence. For LAN applications, C_n^2 can be assumed constant along the path, and

$$\rho_0 = [0.55C_n^2 k_0^2 L^2]^{-3/5} \qquad (5.15)$$

for a spherical wave, and

$$\rho_0 = [1.45C_n^2 k_0^2 L^2]^{-3/5} \qquad (5.16)$$

for a plane wave.

Now, from a statistical point of view, we will consider three types of random perturbations to the signal.

5.4.1 Beam Spread and Beam Wandering

The wavefront distortion is accompanied by random variations in the wave normals, which in turn modify the shape and position of the beam spot at the receiver plane. Here we will separate the random motion of the position of the centroid of the beam (i.e., the energy center of the instantaneous beam-intensity distribution in the receiver plane) from the actual shape of the spot shape. The beam centroid varies with time in a random way, approximately following a Gaussian distribution. This is usually called beam wandering. If we superimpose all of the instantaneous beam shapes, but spatially shifted so as to make the beam centroids coincide at the same point, we obtain as an average a beam spot that is wider than the one expected for free-space diffraction, an effect known as beam spread or beam broadening. For an originally Gaussian beam, the beam spread induces a Gaussian spot. The average beam spot so described is called a short-term beam spot. If we average the beam in a long term so that the effect of beam wander is also included, the angular spread of the beam is related to the loss of coherence. In the far field [30]:

$$\Omega_b^2 = \Omega_0^2 + \left(\frac{\lambda}{\pi\rho_b}\right)^2 \qquad (5.17)$$

where both Ω_b and Ω_0 are the angular apertures defined at $1/e$ of the on-axis intensity for the turbulence-broadened beam, and the free-space beam respectively. The long-term beam spread is related to the diffraction of a Gaussian aperture of radius equal to the coherence radius of the channel as seen from the emitter ρ_b. Here we will call this radius the "backward" coherence radius to indicate that it must be calculated from the receiver to the emitter. As stated previously, in LAN applications C_n^2 is constant along the beam path, and forward and backward coherence radii coincide since, in the far field, the spherical wave approximation is reasonable, and the coherence radius is calculated from equation (5.15).

For the short term, the same expression stands, but ρ_b must be substituted by the "short-term" coherence radius which can be obtained as [31]:

$$\rho_{st} = \rho_b \left[1 + 0.37 \left(\frac{\rho_b}{D_0} \right)^{1/3} \right] \tag{5.18}$$

The statistics of beam wandering are characterized by a Rayleigh distribution with variance

$$\sigma_{\theta_b}^2 = \left(\frac{\lambda}{\pi \rho_b} \right)^2 - \left(\frac{\lambda}{\pi \rho_{st}} \right)^2 \tag{5.19}$$

However, beam wander only occurs in the near field of the "coherent-effective" emitting aperture (i.e., the smallest of the antenna apertures and the "backward" coherence radius ρ_b). Otherwise, the beam is strongly distorted and often breaks in several spots, but there is no apparent "centroid wander" and only the "long-term" (average) beam spread has a physical significance.

An important implication drawn from beam spread is that beam collimation is limited to a saturation value depending on the turbulence strength. This is important, since one of the advantages of lightwaves over microwaves is that very directive links can be obtained with relatively small antenna diameters. As was shown in (5.17), if the free-space beam aperture is much smaller than the beam broadening, this effect dominates the beam width. This saturation in the beam width must be considered in strong turbulent conditions since it is not possible to improve the directivity of the beam by increasing the antenna size further beyond the coherence radius.

5.4.2 Angle-of-Arrival Fluctuations and Focal Spot Broadening

The reciprocal effect of beam wander is the fluctuation of the average wave normal at the receiving antenna aperture. The phase difference across an antenna of diameter D_R when a wave front comes with an angle a_a is approximately

$$\Delta\phi = \phi(0) - \phi(D_R) = kD_R \sin(\alpha) \approx kD_R\alpha \qquad (5.20)$$

Then, the mean square angle of arrival of a distorted wave front is the following [30]:

$$\langle \alpha_a^2 \rangle \approx \frac{\langle \Delta\phi^2 \rangle}{k^2 D_R^2} \approx \frac{D_\phi}{k^2 D_R^2} \qquad (5.21)$$

where D_ϕ is the phase structure function, which can be approximated to the wave structure function D_w for beams that do not scatter light out of the beam path.

Using the relation between wave structure function and coherence radius, the angle-of-arrival fluctuations follow a Rayleigh distribution around the average angle of arrival (which is defined by the beam path to the detector) whose variance is

$$\sigma_{\alpha_a}^2 = \langle \alpha_a^2 \rangle \approx [k^2(D_R \rho_f^5)^{1/3}]^{-1} \qquad (5.22)$$

In long-haul links, where the turbulence structure may vary along the optical path, the coherence radius is calculated in the "forward," that is, propagation, direction. For link distances relevant to LANs, where C_n^2 is constant, $\rho_f = \rho_b = \rho_0$.

The angle of arrival as previously defined corresponds to the "coarse" wavefront structure. The piece of wave front that is intercepted by the aperture is still rugged, and the antenna collects its energy in a irregular spot. As with beam broadening, we can consider that on average the collected spot is Gaussian, taking into account that the Airy spot collected from a coherent plane wave in the focal plane of an antenna with circular aperture can be very accurately approximated as a Gaussian. In the long term, the focal plane spot is spread by both its irregular instantaneous shape, and its position fluctuations are related to the angle of arrival.

The Gaussian angular width, defined at $1/e$ of the on-focus intensity, of the collected spot is:

$$\Omega_{lt}^2 = \left(\frac{2\lambda}{\pi D_r}\right)^2 + \left(\frac{\lambda}{\pi \rho_r}\right)^2 \qquad (5.23)$$

where D_r is the diameter of the receiver antenna. The actual spot size at the focal plane is obtained by multiplying the angular value by the focal length of the optics f_c.

In the short term we will have an instantaneous irregular spot varying in position. In the simplest approach, its average intensity distribution can be obtained by subtracting twice the variance of the angle-of-arrival fluctuations (since the diameter is defined at $1/e$ of the on-axis intensity) to the long-term value. If the collection optics have a circular aperture of diameter D_r, we have:

$$\Omega_c^2 = \left(\frac{\lambda}{\pi \rho_f}\right)^2 \left[1 + \left(\frac{2\rho_f}{D_r}\right)^2 - \frac{1}{2}\left(\frac{\rho_f}{D_r}\right)^{1/2}\right] \tag{5.24}$$

5.4.3 Power Scintillation

To determine the intensity of fluctuations of the received radiation due to turbulence-induced random interference it is useful to use the log amplitude of the field, defined as the natural logarithm of the field amplitude normalized to its average value. The Rytov theory of weak turbulence predicts that for a spherical wave and a homogeneous path, the log amplitude follows a normal distribution with variance [29]:

$$\sigma_\chi^2 = 0.124 C_n^2 k_0^{7/6} L^{11/6} \tag{5.25}$$

for spherical waves, and

$$\sigma_\chi^2 = 0.3075 k_0^{7/6} C_n^2 L^{11/6} \tag{5.26}$$

for plane waves. Contrary to the beam-size effects, it seems to be proven that the effect of scintillation on collimated beams approximate very well to the plane-wave case [30].

However, the Rytov theory does not consider multiple scattering effects, and it cannot be applied to moderately strong turbulences. Experimental measurements show that the variance of the log amplitude does not grow linearly along with the index structure parameter. Instead, it reaches a saturation point, which is shown in Figure 5.5. Entering the estimation of the variance of scintillation determined from the previous expressions in the horizontal axis of this figure, one can obtain the expected value for an experimental situation to be used in the forthcoming calculations.

The fluctuations of the wave intensity follow a log-normal distribution of variance

$$\sigma_I^2 = e^{4\sigma_\chi^2} - 1 \tag{5.27}$$

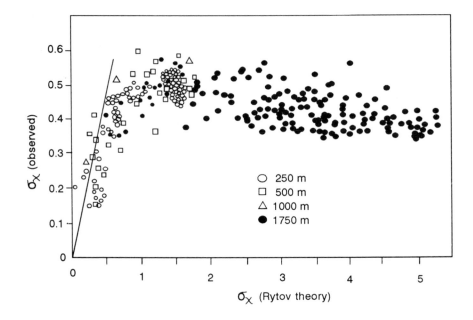

Figure 5.5 Experimental versus theoretical (Rytov model) values of the standard deviation of the log amplitude (after [29], courtesy of the Society of Photo-Optical Instrumentation Engineers).

However, if the coherence radius in the receiver antenna is smaller than its aperture, constructive and destructive interferences may coexist in the detector, averaging its effect. The larger the antenna size, the more the averaging. The calculation of the averaging effect of a turbulence is somehow involved. Fried [32] made calculations based on the Rytov theory for a plane wave travelling on a homogeneous turbulence, which gives a simple expression for the averaging factor in the conditions of small variance, and graphs for stronger scintillations. However, the Rytov theory is not valid for the strong turbulence case, and a first-order correction has been recently proposed to generalize the results to the multiple scattering regime [28]. First of all we will notice that, for a perfectly correlated scintillation along the receiver aperture, the collected power variance would be given by

$$\sigma_{P_{coh}}^2 = \pi \left(\frac{D_R}{2}\right)^2 \sigma_I^2 \tag{5.28}$$

To account for the antenna averaging effect, this amount should be reduced in the following way:

$$\sigma_p^2 = \frac{\pi(D_R/2)^2\sigma_I^2}{1 + (D_R/2)^2(k_0/L + 1/\rho_0^2)} \tag{5.29}$$

where ρ_0 is given by (5.16).

Notice that $(k_0/L)^{1/2}$ gives the transverse correlation radius of intensity fluctuations in weak turbulence and is independent of the turbulence strength. Thus, increasing the antenna aperture indefinitely does not increase indefinitely the averaging of the fluctuations in the collected power. On the contrary, this effect is asymptotically limited by a saturation state whose variance would be a constant value in weak turbulence, but depends on the turbulence strength for very strong turbulences. It is useful to define the antenna averaging factor as

$$\Theta(D_R) = \frac{\sigma_p^2}{\sigma_I^2} = \left[\frac{4}{\pi D_R^2} + \frac{2}{\lambda L} + \frac{1}{\pi\rho_0^2}\right]^{-1} \tag{5.30}$$

which can be seen as the effective coherent-antenna aperture, obtained from the "convolution" of the actual antenna area, the weak scintillation coherent area, and the strong turbulence coherent area.

It has been experimentally proven that, except under too-strong turbulences, the power fluctuations also follow a log-normal distribution. Thus, it is convenient to define a power log-amplitude χ_p as one-half of the natural logarithm of the instantaneous power divided by its average value. This random variable follows a normal distribution whose mean equals minus its variance. In this way, and following (5.27) and (5.30), the antenna averaging effect can be expressed by

$$e^{4\sigma_{\chi_p}^2} - 1 = \Theta(D_R)[e^{4\sigma_\chi^2} - 1] \tag{5.31}$$

where $\sigma_{\chi_p}^2$—the variance of the power log amplitude—can be calculated.

5.5 STEADY POWER BUDGET IN ATMOSPHERIC OPTICAL LINKS

Since the atmosphere is a random channel, we will divide the analysis of its characteristics into two parts. First we will deal with the average properties so that we can follow a typical scheme for the power budget, as for any communications link. Later we will perform a statistical analysis which will give us the probability distribution that the received power oscillates around this value. Since atmospheric inhomogeneities inducing such oscillations are slow compared to the transmission rates, the occasional fading of power in the receiver link will be associated with burst errors.

Concerning the average power at the detector, many technical papers and textbooks have dealt with the problem. For our convenience, we will present the budget equation such that a link margin (*LM*) is defined as the "average power in excess of the emitter," or alternatively the "average losses in excess allowed for the channel," in the following way:

$$LM = P_e - L_{topt} - L_{prp} - L_{point} - L_{atm} - L_{ropt} - L_{col} - P_r - SM \quad (5.32)$$

where P_e and P_r are, respectively, the emitted power and the receiver sensitivity, both in relative units (dBm or dBW) and all the other terms are the losses expressed in dB (described below).

The receiver sensitivity is defined as the minimum received power necessary to keep the signal-to-noise ratio (SNR) above a value that defines the link quality. The noise level in the receiver and the SNR determine the minimum detectable power. For digital transmission, the link quality is defined in terms of the bit error rate (BER), and its relation to the SNR depends on the type of detector and the transmission code [33]. This relation can be found in specialized books on optical communications. A schematic of the link budget can be seen in Figure 5.6. The system margin (SM) must account for all of the possible degradations of the system

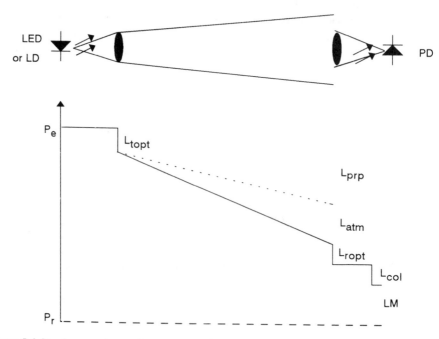

Figure 5.6 Steady power losses along an atmospheric IR path.

during its lifetime and is especially important for optoelectronic emitters, as indicated in CCITT recommendations for line systems.

5.5.1 Telescope Losses

Losses in the optics of both transmitter and receiver telescopes have been respectively designed as L_{topt} and L_{ropt}. Optical elements in the antennas include collimating elements (either lenses or mirrors); receiver optics usually include an interference filter to protect the receiver from excess background radiation, and sometimes beam splitters are incorporated, for instance for emission and reception-path combinations in the same antenna. Telescope losses account for both Fresnel reflections and glass (or mirror) absorption losses. Here, losses are considered to be those of ideal devices from a geometric point of view, and can be calculated directly from classic optics formulae and data sheets of glassoptics manufacturers. Regarding the Fresnel losses of the lenses, for instance, since all of them will probably have similar antireflective coatings, they may be computed as

$$-20N \log(T_f) \tag{5.33}$$

T_f being the Fresnel transmission coefficient for power of a single interface, and N the number of optical elements, each of them presenting two interfaces. Losses in the glass must account for the lens thicknesses and glass transmission. Losses in interference (or any other) filters are directly provided by manufacturers, and for losses in beam splitters it is necessary to consider its angle with respect to the beam path.

5.5.2 Propagation Losses (L_{prp})

Propagation losses in the vacuum can be calculated from diffraction of the emitter aperture and the area of the receiver antenna. Usually they are split in the difference between the propagation losses for an isotropic source at a distance L and the antenna gains. In the atmospheric channel there is an additional penalty due to the average beam broadening induced by the turbulence:

$$-L_{prp} = G_t - L_{isot} - L_{turb} + G_r \tag{5.34}$$

where $L_{isot} = 10 \log[\{4\pi L/\lambda\}^2]$

Since the field at a receiver aperture of active area A_r can be approximated to a plane wave, the receiver antenna gain for any geometry is

$$G_r = 10 \log[4\pi A_r/\lambda^2]$$

The beam diffracted from the transmitter antenna can be usually approximated by a Gaussian. This is exact for a laser beam and very accurate for an Airy beam (i.e., a beam emitted from a circular aperture uniformly illuminated). Since the distortion of the beam by the turbulence follows Gaussian statistics, it is convenient to write the emitter antenna gain G_t as a function of an effective Gaussian beam divergence, which is the divergence of the Gaussian that best fits the actual emitted beam:

$$G_t = 10 \log\left(\frac{2}{\Omega_0}\right)^2 \qquad (5.35)$$

for Ω_0 defined as the angle at which the beam irradiance is $1/e$ times the value at the beam axis (i.e., the rms divergence of the electric field distribution). For a Gaussian beam with a radius at the antenna aperture defined at $1/e$ of the on-axis irradiance R_o, $\Omega_0 = \lambda/(2\pi R_o)$, while the effective Gaussian divergence for an Airy beam emitted from a circular aperture of diameter D_t is $\Omega_0 = 2\lambda/(\pi D_t)$.

The rms beam broadening comes, given from expression (5.17) , as $\Omega_{\text{turb}} = \lambda/\pi\rho_b$. If the receiver is in the "coherent near field" of the source, then the short-term coherence radius—calculated from (5.18)—must be used, and a beamwandering effect must be considered in the statistical budget. The additional broadening of the beam introduces a penalty in the propagation losses, which may be written as:

$$L_{\text{turb}} = 10 \log[1 + (\Omega_0/\Omega_{\text{turb}})^2] \qquad (5.36)$$

5.5.3 Mispointing Losses (L_{point})

The propagation losses presented above are calculated on the basis that the beam axis reaches the receiver antenna. Usually antenna pointing can be done while setting the link, optimizing the average received power after a coarse visual pointing. This must be done carefully due to the random character of the signal level in the receiver, which has been indicated in Section 5.4. However, even if pointing has been very accurate, this situation may vary with time. Seasonal or day-to-night variations due to beam refraction are not going to be noticeable in the distances relevant to AOLANs, but the effect of vibrations can be important. In most situations the optical terminals will be placed in the upper part of the buildings where the vibrations are more noticeable, especially in urban areas with heavy traffic or in seismic regions.

If there is some mispointing, the collected irradiance is below the calculated value. For the Gaussian approximation of the beam, the ratio between the beam intensity for a mispointing angle e and the on-axis beam intensity is

$$L_{\text{point}} = -10 \log[I(\epsilon)/I(0)]$$
$$= -10 \log[e^{-(\epsilon/\Omega_b)^2}] \qquad (5.37)$$
$$= 4.3429(\epsilon/\Omega_b)^2$$

The factor 4.3429 converts nepers into decibels, and can be omitted if the budget is being performed in nepers.

5.5.4 Atmospheric Losses (L_{atm})

The atmospheric losses are the result of the attenuation of atmospheric gases and aerosols, therefore calculated as αL for a total path length L. The attenuation coefficient α is given by (5.1).

5.5.5 Collection Losses (L_{col})

Collection losses may appear when part of the radiation is focused by the receiver optics out of the detector due to the size or position of the focused spot. The position depends on the wavefront angle of arrival, which is a random variable in turbulent media and will be considered in the statistical budget. The spot is approximately Gaussian and, in angular terms is given by (5.24), and the corresponding losses are:

$$L_{\text{col}} = -10 \log[1 - e^{-(\text{fov}/\Omega_c)^2}] \qquad (5.38)$$

where fov is the angular field of view of the detector.

5.6 STATISTICAL BUDGET FOR ATMOSPHERIC OPTICAL LINKS

Turbulence is responsible for random variations of the received power around the steady value calculated in the previous section. It may happen that an "instantaneous" analysis of the link would give a negative power budget at some moments. In these periods, the received power would be below the receiver sensitivity (i.e., the BER will increase above its standard value). Since turbulence spectrum is lowpass, the fluctuations on the collected power will be slow compared to the bit rate, and the periods of time when the BER is high are known as "burst errors." The statistics of the perturbations induced by the turbulences will allow us to determine the probability of such situations, which is usually known as probability of burst error (PBE). The quality of the communications is better described by the percent of time that the link is available than by the average BER. An atmospheric channel therefore, should be described by three conditions:

1. The maximum BER, which defines the quality of the link under "proper operation" conditions.
2. The limiting atmospheric attenuation, which produces a complete interruption of the beam (i.e., an error rate above that required for good transmission). For a good design, the meteorological conditions of the area will provide a tolerable atmospheric transmission during most of the time.
3. The PBE that the link quality is degraded during short bursts under "clear transmission" conditions.

The PBE is defined in terms of the probability density function (pdf) of random losses. These losses come from the three effects presented below: beam wander, angle-of-arrival fluctuations, and (mostly) power scintillation. For the sake of simplifying the expressions involved, we will express losses in nepers, and their origin will be respectively represented as l_{wan}, l_{arr}, and l_{sci}. The pdf of each of them will be named by an f with the respective suffix. Losses in dB will be represented by a capital L with the corresponding suffix, and they are related to losses in nepers by the simple expression:

$$L = 10 \times \log(e) \times l \qquad (5.39)$$
$$= 4.3429 \times l$$

where log expresses the decimal logarithm, and e is the basis of natural logarithms.

The pdf of the sum of two independent random variables is given by the convolution of each independent pdf [34]. Thus, the total losses are

$$l = l_{wan} + l_{arr} + l_{sci} \qquad (5.40)$$

and its pdf

$$f(l) = f_{wan}(l)*f_{arr}(l)*f_{sci}(l) \qquad (5.41)$$

The PBE is given by the probability that losses are larger than the link margin (LM) (see Fig. 5.7):

$$PBE = \int_{lm}^{\infty} f(l)dl \qquad (5.42)$$

Each of the components of the pdf of losses are described below.

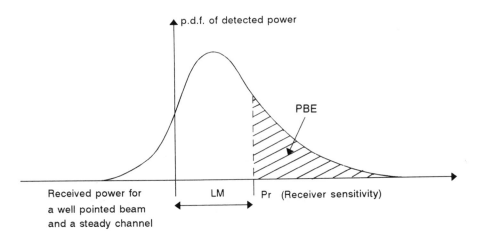

Figure 5.7 Illustration of the calculation of the PBE from the pdf of received power.

5.6.1 Probability Density Function of Losses Due to Beam Wandering

Beam wander is noticeable when the receiver is placed in the coherent near field of the emitter. As was indicated, this is the near field of the "coherent" aperture size of the transmitter, defined as the smallest of the emitter-antenna aperture, and the coherence radius of the beam distorted by turbulences. The beam-wander losses need not be accounted for if the link distance is larger than $\min(k_0\rho^2, k_0D_T^2/4)$, where D_T is the diameter of the transmitter aperture.

When present, the beam wander presents a Rayleigh distribution of variance defined in (5.19)

$$f_{\text{wan}}(\theta_b) = (\theta_b/\sigma_{\theta_b}^2)e^{-\theta_b^2/(2\sigma_{\theta_b}^2)} \tag{5.43}$$

When the beam wandering is relevant, the short-term beam divergence is Ω_{st} calculated from (5.17) by substituting ρ_0 by ρ_{st} calculated from (5.18). The mispointing losses for a Gaussian beam with an angular divergence Ω defined at $1/e$ of the on-axis intensity are now:

$$l_{\text{wan}}(\theta_b) = -\ln(I(\theta_b)/I(0)) = (\theta_b/\Omega_{\text{st}})^2 \tag{5.44}$$

Notice that the inverse function giving the beam direction as a function of losses involves a square root, which is a doublevalued function. However, the restriction that the Rayleigh distribution is only defined for positive values of the variable disregards the negative solution. Therefore, the pdf of the beam direction-associated losses is [35]

$$f_{\text{wan}}(l_{\text{wan}}) = \frac{1}{2}(\Omega_{\text{st}}/\sigma_{\theta_b})^2 e^{-1/2(\Omega_{\text{st}}/\sigma_{\theta_b})^2 l_{\text{wan}}} \qquad l_{\text{wan}} > 0 \tag{5.45}$$

which is an exponential distribution defined only for positive values of the losses. The width of the distribution is given by the ratio $\sigma_\theta/\Omega_{\text{st}}$ between the rms angular fluctuations of the beam direction and the short-term broadened beam divergence.

5.6.2 Probability Density Function of Losses Due to Angle-of-Arrival Fluctuations

Angle-of-arrival fluctuations follow a Rayleigh distribution (the same as for beam wandering) whose variance is now given by (5.22):

$$f_{\text{wan}}(\alpha_a) = (\alpha_a/\sigma_{\alpha_a}^2)e^{-\theta_b^2/(2\sigma_{\alpha_a}^2)} \tag{5.46}$$

Due to angle-of-arrival fluctuations, part of the focal spot in the receiver optics may occasionally fall out of the detector. The collection losses, which in (5.38) were calculated for a centered beam, will actually depend on the angle of arrival and represent the integral of a two-dimensional Gaussian on a miscentered circular area, as illustrated in Figure 5.8, and are given by [35]:

$$l_{\text{arr}}(\alpha_a) = -\ln[(2e^{-(\alpha_a/\Omega_c)^2}/\Omega_c^2)\int_0^{\text{FOV}} \phi e^{-(\phi/\Omega_c)^2}I_0(2\alpha_a\phi/\Omega_c^2)d\phi] - l_{\text{col}} \tag{5.47}$$

where l_{col} are the collection losses at angle of arrival zero, expressed in nepers and I_0 is the modified Bessel function of the first kind and order zero.

Since (5.47) is not invertible analytically, the distribution function of losses due to angle of arrival $f_{\text{arr}}(l_{\text{arr}})$ must be generated numerically. Figure 5.9 shows the Rayleigh distribution of angle of arrival and the probability density distribution of its associated losses for a particular atmosphere (see captions for numerical details).

5.6.3 Probability Density Function of Scintillation Losses

The scintillation effect was described by the statistics of the power log amplitude. The fact that the optical-power fluctuations present a log-normal distribution

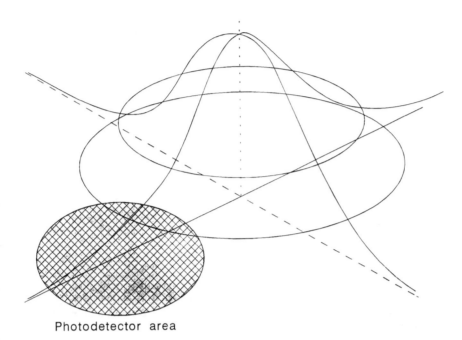

Photodetector area

Figure 5.8 Collected radiation spot distribution on the focal plane when the wavefront angle of arrival is not orthogonal to the receiver optics.

requires that its mean value equal its variance. Thus, the averaged power log amplitude is a Gaussian distribution

$$f(\chi_p) = (1/\sqrt{2\pi}\sigma_{\chi_p})e^{-(1/2)[(\chi_p + \sigma_{\chi_p}^2)/\sigma_{\chi_p}]^2} \tag{5.48}$$

The scintillation losses can be defined as

$$l_{sci} = -\ln(P/\langle P\rangle) = -2\chi_p \tag{5.49}$$

Therefore, the probability density function of the scintillation losses is also a Gaussian distribution with mean σ_χ^2 and standard deviation $2\sigma_\chi$ [36]:

$$f_{sci}(l_{sci}) = (1/2\sqrt{2\pi}\sigma_{\chi_p}) \exp(-[(l_{sci}/2\sigma_{\chi_p}) + \sigma_{\chi_p}]^2/2) \tag{5.50}$$

Notice that, different from both of the random losses described previously, scintillation losses can be negative. This means that the received optical power is

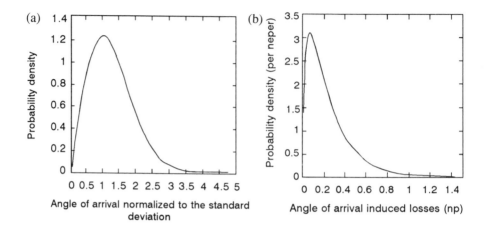

Figure 5.9 Comparison between (a) the Rayleigh distribution function of the angle-of-arrival fluctuations and (b) distribution function of its associated losses calculated for a turbulence characterized by a coherence radius of 0.95 cm and a receiver consisting of a circular photodetector of 2-mm diameter placed at the focal distance (5.4 cm) of a lens with a diameter of 17 cm. Collection losses (for zero AOA) in this case are 0.12 nepers.

above its average value due to a constructive interference of the distorted wave front on the detector.

5.7 CONCLUSIONS

Infrared links in open air are affected by a variety of meteorological perturbations that have no effect on usual radioelectric links. These effects have been described as inherent features of the atmospheric channel, which is a dynamic one. All of the effects of the atmosphere have been quantified as follows.

On the one hand, formulae or graphs are provided to estimate its contribution to the link losses. LOWTRAN is the most complete one, and it can be used with broad-spectrum light sources.

On the other, emphasis was put on the effects of the turbulences, which have been quantified on the basis of the coherence radius, a parameter that fully characterizes the turbulence. The coherence radius depends on the optical-field distribution, and, at relevant distances, narrow beams can be represented by spherical waves when computing scintillation, beam spread, beam wander, or wavefront angle-of-arrival fluctuations. The radius of coherence depends on one meteorological parameter: the refractive index structure parameter. Its value may be available for a variety of places and meteorological conditions since it is traditionally used to define the quality of seeing in astronomic observation. Nevertheless, typical values are provided in any case.

Atmospheric attenuation and coherence radius are, therefore, the necessary parameters to carry on the link evaluation. The link design is done on the basis of the desired link quality. We have suggested that the analysis of the link performance be split into two parts.

First, a steady power budget establishes which is the most probable received power of a certain link. A certain safety margin (received power to receiver sensitivity) must be available for random losses around this value. As for any digital link, the receiver sensitivity depends on the desired BER. Atmospheric effects impacting this budget are the total attenuation and the beam broadening.

Random losses will exceed this margin at given intervals where the received power will be below the receiver sensitivity and the BER will be higher than the specified. Since random variations of turbulence are slow compared to the bit rate, these intervals will result in burst errors. The second part of the analysis is a statistical budget of all of the lossinducing turbulence effects on the beam. The PBE is calculated as the probability that random losses exceed the link margin, and the distribution function of random losses is obtained from the convolution of the individual distribution functions of losses due to each of the effects, which can always be obtained from physical data on the optics of the terminals, and the coherence radius of the beam after the distorting effect of turbulences.

The quality of the link is here defined by the maximum BER under "normal operation conditions," and the PBE, which defines the probability to be in "abnormal operation conditions" (i.e., (1 PBE) gives the percent of time in which the link is operational with standard transmission quality). Both parameters give a more complete picture of the link than just specifying the average BER. The link budget analysis presented here is a simplified version of a more comprehensive model developed for the evaluation of satellite-to-Earth optical links where terminals are provided with automatic pointing and tracking systems [35,37].

REFERENCES

[1] López, F. J., P. Menéndez-Valdés, R. Beltrán, M. J. Betancor, J. M. Otón, and M. López-Amo. "Red local híbrida para la Universidad Politécnica de Madrid," *Actas II Jornadas de Comunicaciones Opticas*, Madrid, June 2–4, 1985, pp. 40–43.

[2] Betancor, M. J., R. García, A. Julián, M. López-Amo, F. J. López, and J. M. Otón. "Red híbrida de Comunicaciones Opticas para la Universidad Politécnica de Madrid," *Mundo Electrúnico*, No. 192, 1989, pp. 51–55.

[3] Binder, B. T., P. T. Yu, J. H. Shapiro, and J. K. Bounds. "An Atmospheric Optical Ring Network," *IEEE Trans. on Communications*, Vol. 38, 1990, pp. 74–81.

[4] Zuev, V. E. *Laser Beams in the Atmosphere*, Consultants Bureau, New York, 1982.

[5] Wolfe, W. L., and G. J. Zissis, eds. *The Infrared Handbook*, 3rd printing, Ann Arbor: ERIM-SPIE, 1989.

[6] Zuev, V. E. *Propagation of Visible and Infrared Radiation in the Atmosphere*, New York: John Wiley & Sons, 1974.

[7] Brown, T. L., and H. E. Le May, Jr. *Chemistry, The Central Science*, 2d ed. Englewood Cliffs: Prentice Hall, Inc., 1981.

[8] McEwan, M. J., and L. F. Phillips. *Chemistry of the Atmosphere*, London: Edward Arnold Publishers Ltd., 1975.

[9] Goody, R. M. *Atmospheric Radiation Vol. 1: Theoretical Basis*, Oxford: Oxford University Press, 1964.

[10] Rothman, L. S., A. Goldman, J. R. Gillis, R. R. Gamache, H. M. Pickett, R. L. Poynter, N. Husson, and A. Chedin. *Applied Optics*, Vol. 22, 1983, p. 1616.

[11] Rothman, L. S., R. R. Gamache, A. Barbe, A. Goldman, J. R. Gillis, L. R. Brown, R. A. Toth, J.-M. Flaud, and C. Camy-Peyret. *Applied Optics*, Vol. 22, 1983, p. 2247.

[12] Kneizys, F. X., G. P. Anderson, E. P. Shettle, W. O. Gallery, L. W. Abreu, J.E.A. Selby, J. H. Chetwind, and S. A. Clough. "Users Guide to Lowtran 7," AFGL-TR-88-0177, ERP No 1010, Air Force Geophysics Laboratory, Hanscom AFB (Massachusetts), 1988.

[13] Menéndez-Valdés, P. "Atmospheric Transmission and Climatic Effects" in *Assessment of Atmospheric Losses on an Optical Link Budget*, Final Report to ESA Contract No. 8131/88/NL/DG, UPC, UPM, IAC, ONERA, Barcelona, October 1989.

[14] Mie, G. "Beiträge zur Optik trüber Medien, speziell kolloidaler Metallösungen," *Ann. der Physik*, Vol. 25, 1908, pp. 376–445.

[15] Born, M., and E. Wolf. *Principles of Optics*, 6th ed. Oxford: Pergamon Press, 1980.

[16] McCartney, E. J. *Optics of the Atmosphere*, New York: John Wiley & Sons, 1976.

[17] Felske, J. D., Z. Z. Chu, and J. C. Ku. "Mie scattering subroutines (DBMIE and MIEV0): a comparison of computational times," *Applied Optics*, Vol. 22, No. 5, August 1, 1983, pp. 2240–2241.

[18] Nilsson, B. "Meteorological influence on aerosol extinction in the 0.2-40 mm wavelength range," *Applied Optics*, Vol. 18, No. 20, 1979, pp. 3457–3473.

[19] Shettle, E. P., and R. W. Fenn. "Models of the atmospheric aerosols and their optical properties," in AGARD-Conference Proceedings No. 183 *Optical Propagation in the Atmosphere*, Lingby (Denmark), October 27–31, 1975, pp. 2-1 to 2-16.

[20] Gathman, S. G. "Optical properties of the marine aerosol as predicted by the Navy aerosol model," *Optical Engineering*, Vol. 22, 1983, pp. 57–

[21] J. M. Rosen, D. J. Hofmann. "Optical modeling of stratospheric aerosols: present status," *Applied Optics*, Vol. 25, 1986, pp. 410–419.

[22] Menéndez-Valdés, P. "Near-infrared attenuation of Saharan dust at the Tenerife and La Palma Observatories," *Applied Optics*, Vol. 31, 1992, pp. 457–459.

[23] Blanco, F., and P. Menéndez-Valdés. "A weather dependent model of coastal fog attenuation," *Optical Engineering*, in press.

[24] Kneizys, F. X., E. P. Shettle, W. O. Gallery, J. H. Chetwynd, L. W. Abreu, J.E.A. Selby, S. A. Clough, and R. W. Fenn. "Atmospheric Transmittance/Radiance: Computer Code LOWTRAN 6," AFGL-TR-830187, ERP No. 846, Air Force Geophysics Laboratories, Hanscom AFB (Massachusetts), 1983.

[25] Joss, J., and A. Waldvogel. "Raindrop size distributions and sampling errors," *J. Atmos. Sci.* No. 26, 1971, pp. 566–569.

[26] Sekhon, R. S., and R. C. Srivasteva. "Doppler radar observations of drop-size distributions in a thunderstorm," *J. Atmos. Sci.* No. 28, 1971, pp. 983–984.

[27] Tatarski, V. *Wave Propagation in a Turbulent Medium*, New York: Dover Publishing, Inc., 1961.

[28] Karp, S., S. M. Gagliardi, S. E. Moran, and L. B. Sttots. *Optical Channels*, New York: Plenum Press, 1988.

[29] Hufnagel, R. E. "Propagation through atmospheric turbulence," in *The Infrared Handbook*, edited by W. L. Wolfe and G. J. Zissis, IRIA-ERIM, and SPIE, 3rd printing, 1989.

[30] Fante, R. L. "Electromagnetic beam propagation in turbulent media," *Proc. IEEE,* Vol. 63, 1975, pp. 1669–1691.

[31] Yura, H. T. "Short-term average optical-beam spread in a turbulent medium," *J. Opt. Soc. Am.,* Vol. 63, 1973, pp. 567–572.

[32] Fried, D. L. "Aperture Averaging of Scintillation," *J. Opt. Soc. Am.*, Vol. 57. No. 2, February 1969, pp. 169–175.

[33] Personik, S. D. "Receiver design for digital fiber optical communication system," *Bell Systems Technical Journal*, Vol. 50, 1973, pp. 843–886.

[34] Papoulis. *Probability, Random Variables and Stochastic Processes*, 2d ed., Singapore: McGraw-Hill, 1984.

[35] Menéndez-Valdés, P., and E. Fernández. "Link budget model and applications for laser communications through the atmosphere," *Proc. SPIE,* Vol. 1866, in press (1993).

[36] Scott, P. W., and P. W. Young. "Impact of temporal fluctuations of signal-to-noise ratio (burst error) on free space communications system design," *Proc. SPIE,* Vol. 616, 1986, pp. 174–181.

[37] Menéndez-Valdés, P. "Atmospheric model and link budget definition," *WP3300 Task Report,* UPM, Madrid, January 1991; and , "Link budget model for an atmospheric link provided with tracking," *UPM Technical Note,* No. 4, Addendum to WP3300 Report, UPM, Madrid, July 1991. ESA contract No. 8973/90/NL/SG.

Chapter 6
Properties of Wireless RF Channels

Philip Constantinou
National Technical University of Athens
Dept. of Electrical Engineering, Electroscience Division
Patission 42, Ave. 10682 Athens, Greece

6.1 PROPAGATION MODELS

6.1.1 Definition of a Mobile Channel Model

A mobile channel model is a set of mathematical expressions into which channel characteristics obtained from field measurements can be inserted. This is done to predict the potential performance of a proposed mobile communication system.

The basis for a model may be either theoretical or empirical, or a combination of these two. Theoretical propagation models allow recognition of the fundamental relationships that apply over a broad range of circumstances. They also allow definition of relationships that exist among any combination of input parameters. Empirical models are derived from measurements and observation and offer a major advantage in that all environmental influences are implicit in the result regardless of whether or not they can be separately recognized and theoretically studied. Empirical models offer the opportunity to provide probabilistic descriptions of the propagation phenomena. The validity of empirical models is limited only by the accuracy with which individual measurements are made and by the extent to which the environment of the measurements adequately represents the physical environment in which the model is to be applied.

6.1.2 Propagation Aspects

In a mobile channel the transmitted waves are scattered, diffracted, and attenuated. The various scattered components interfere, building up irregular fields distribution. The signal at the receiver is therefore attenuated and distorted. The severity of this disturbance depends on the specific physical properties of the scattering environment. The propagation of electromagnetic waves either near the ground or inside a building due to diffraction, scattering, reflection, and absorption of the incoming signal is broken into several components that are attenuated and delayed differently.

The signal at the receiver antenna is thus composed of a direct component and a delayed, scattered component. The direct path can be obstructed, depending on the antenna location and shadowing conditions. The degree of shadowing varies very strongly with the movement of the mobile antenna, leading to equivalent time fluctuations of the received power of the direct and delayed components.

Due to the multipath propagation, the received signal will arrive at the receiver from different directions, with different path lengths and with different time of arrival. Therefore if an impulse is transmitted, by the time this impulse will arrive at the receiver it is no longer an impulse but a pulse with a spread width which is called the delay spread. This delay spread will limit the maximum transmission rate of a digital signal.

The different time delays in two signals closely spaced in frequency can cause the two signals to become correlated. The bandwidth in which the amplitudes or the phases of two received signals are strongly correlated is defined as the coherence bandwidth.

As a result of the multipath propagation, the received signal presents rapid fluctuations that are characterized as fast fadings or Rayleigh fadings. The median values of the received signal strength also fluctuates due to large-scale variations along the path. The median value fadings are defined as slow fades or log-normal fading. When the received signal also includes the line-of-sight component, the envelope is Rice-distributed.

From the above it can be seen that the mobile radio channels are randomly dispersive in both time and frequency. Therefore for digital transmission, if the bandwidth of the transmitted symbol is close to the coherence bandwidth of the channel, an increase in the probability of error results from the frequency-selective fading due to multipath propagation. On such channels, therefore, there exists an optimum combination of digital symbol rate and symbol shape that can be chosen to minimize the probability of error. Even if the time-bandwidth product of transmitted symbols is chosen so as to make the channel nonselective, flat fading causes an effective signal-to-noise (S/N) degradation that must be accounted for in the system design.

It is clear that the accurate characterization and modeling of a mobile channel is a necessity. The required information for this characterization includes knowledge of the statistics for signal fading, knowledge of the dispersive nature of the medium, and knowledge of the additive noise that can be expected in a mobile-radio environment.

6.1.3 Modeling Requirements

To characterize the mobile channels requires a complete knowledge of the propagation parameters described above for all environments where the system will operate. Conducting measurements to obtain all propagation parameters for all possible environments is an impossible task and for a limited number of environments a time-consuming exercise. In addition, testing a new system requires repetition of the measurements with the same propagation medium (i.e., a stable propagation environment). Therefore a propagation model is required that provides all the parameters which characterize the mobile channel.

Complete channel characterization is required by the equipment and the mobile-radio systems design engineer. Propagation models that apply to a wide variety of locations, but in a limited frequency band and for limited distances, are needed for general system design, such as when systems are being developed that will operate in many locations. When a given performance objective is to be met in a known location, the specific system design requires a propagation model that accounts for relevant environmental and topographical information.

Allowance for such phenomena can only be made if a model is available that accurately describes the expected probability distribution of the envelope of signals received over a mobile-radio link and the amount of dispersion that can be expected.

6.1.4 Scope of Modeling

In a mobile-communication system, as in any communication system, we can identify three major parts: the transmitter, the channel, and the receiver. At the transmit site, the signal must be conditioned in such a manner so as to overcome distortions that will be introduced by the channel. This task involves coding, pulse shaping, and modulation. Therefore, a priori information about the channel is a necessity. The receiver task is to extract the information signal from the received signal by implementing all the available techniques, such as equalization, demodulation, error correction, error detection, and decoding. These tasks require a complete knowledge of the mobile channel.

In a mobile system, while the receiver is moving it will receive the signal(s) from a channel that continuously changes. This channel variability is a function of time and location of the mobile in each environment and demands, from the design point of view, that certain problems of handover be resolved.

Handover study can be completed assuming knowledge on signal level as a function of the environment, cochannel interference, statistical distribution, and spatial correlation of the fast and slow fading signals. This is a description of the channel(s) where the handover process will take place is required.

The distinctive challenges of designing an equalizer for fading channels are:

1. The necessity for continual and rapid updating of the equalizer coefficients to track the changes in the channel.
2. A need to compensate for multiple paths with nearly equal strengths and for deep nulls within the bandwidth of interest.

To meet the challenge of designing an equalizer, information on tap-weight values, types of fadings experienced, and probability distribution of delay spread must be available to design engineers so that they can determine for what percentage of locations an equalizer, of whatever dimension, has an effect. Error control requires information on CNR and CIR, time dispersion, frequency, and distribution of errors. The types of antennas to be used and power requirements demand information on signal level, interference environment, and noise environment. A knowledge of channel characteristics is necessary so that frequency allocation, channel bandwidths, and channel spacings must be determined.

6.1.5 Channel Parameters

The parameters of the mobile channel that are required for the development of the propagation models are:

1. Signal attenuation-path loss.
2. Signal fading (fast and slow fading).
3. Fading depth and rate of occurrence.
4. Impulse response:
   a. Mean delay,
   b. Delay spread,
   c. Delay interval,
   d. Delay window, and
   e. Channel tap setting and weight values.
5. Noise environment.
6. Interference environment.

6.2 INDOOR MODEL

6.2.1 Path loss

Indoor propagation is a very complex and difficult radio propagation environment because the shortest direct path between transmit and receive locations is usually blocked by walls, ceilings, or other objects. Signals propagate along corridors and other open areas, depending on the structure of the building. The results of measurements [1] indicate that the signal variation inside buildings is approximately Rayleigh distributed for nonline of sight (NLOS) cases, whereas a Rice distribution fits in the case of line of sight (LOS). Therefore, in modeling indoor propagation the following parameters must be considered:

1. Construction materials:
   a. Reinforced concrete,
   b. Brick,
   c. Metal,
   d. Glass,
   e. Wood, and
   f. Plastic.
2. Types of interiors:
   a. Rooms with windows,
   b. Rooms without windows,
   c. Hallways with doors,
   d. Hallways without doors,
   e. Large hallways or open areas,
   f. Corridors with corners, and
   g. Curved corridors.
3. Locations within a building:
   a. Ground floor,
   b. nth floor, and
   c. Basement or garage.
4. Location of transmit (Tx) and receive (Rx) antennas:
   a. Tx and Rx antennas on the same floor,
   b. Tx and Rx antennas within the building, but on different floors, and
   c. Tx antenna outside the building, Rx antenna inside the building.

To assist the user in implementing the model, general indoor-propagation environments are presented in Figures 6.1 and 6.2, with several links between the base station (transmit antenna—B) and the mobile station (receive antenna—M) marked on them. The user of the model can select the link or combinations of links that best satisfies the individual case.

6.2.2 Location: Tx and Rx On the Same Floor

Case 1: B1 to M1 (Figure 6.1). Inside a hall or a large area, the path loss is given by the conventional distance power loss as follows:

$$P_1 = S + 10 \, n_1 \log(d) \qquad (6.1)$$

where P_1 = path loss in dB and S = 37 dB. For f = 1.7 GHz, this constant refers to the 1m intercept point. It is calculated assuming free-space loss (i.e., n = 2):

$$S = 10 \, n \log(4\pi * 1m/\lambda)$$

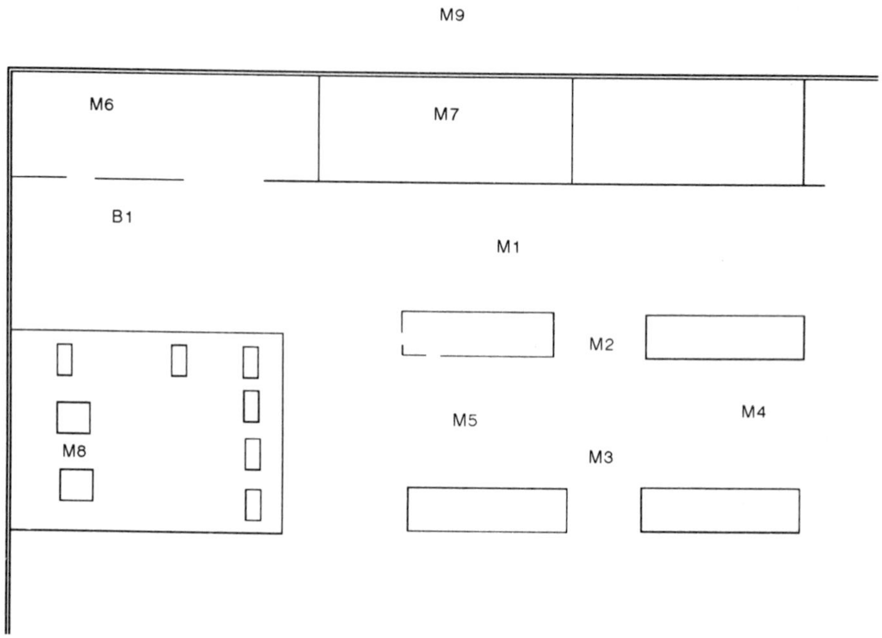

M9

| M6 | M7 | |

B1

M1

M2

M4

M5

M3

M8

Figure 6.1 Indoor environment.

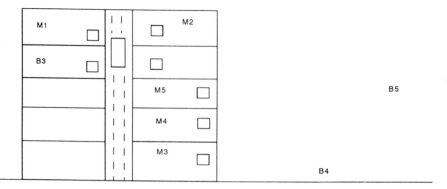

Figure 6.2 Tx and Rx at different floors. B3 inside the building. B4 outside the building, at street level. B5 outside the building, above street level.

n_1, the exponent, depends on the propagation environment. In free space, $n = 2$. For $n_1 < 2$ we have a waveguide effect [2]. Values of n_1 range from 2 to 6. Preliminary measurements conducted by [3] gives the value of $n_1 = 2.9$. The distance between transmitter and receiver, d, is given in meters.

Measurements conducted in Lund, Sweden [4] in three different types of indoor environments conclude that the distance-dependent mean path loss, in the case where LOS exists, is given by

$$M(d) = 10 * 2.9 * \log(d) - 4.5 \text{ (dB)} \qquad (6.2)$$

where d is the distance, in meters, between the reference point S and the receiver.

It is reported in [5] that when a wave encounters an obstacle there will be reflections that can be guided by certain wall configurations and which bring about significantly higher signal levels than would be the case with free-space propagation. This effect has been observed in corridors and in large rooms when there was no obstruction between antennas.

To be able to obtain the value of n, the data from measurements must be presented in the form of a graph relating the mean signal strength as a function of distance in meters. The best-fit line must be presented with the values of S and n. However, one must be cautious about the curve-fitting process. As reported in [3], the value of n is strongly dependent on the portion if data used. For example, if the signal strength values for the first 10m are removed, the least-square method will yield a value of n equal to 4.5 instead of 2.8. Therefore, the curve-fitting process must be described with a parameter indicating the goodness of fit.

Case 2: B1 to M7, M8, and M9 (Figure 6.1). If a wall(s) exists between the transmitter and receiver and the only signal path is through the wall(s), with no channeling effect around corners between Tx and Rx, then path loss is given by:

$$P_2 = S + 10 \, n_1 \log(d_1) + \Sigma L_w \qquad (6.3)$$

where P_2 is the path loss in dB. The exponent n_1 depends on the environment outside the wall. Values of n_1 are defined in case 1, d_1 is the distance between transmitter and external surface of the wall, and L_w is the penetration loss due to the wall(s).

The parameter L_w depends on the type of wall construction between the transmitter and the receiver and the angle of incidence of the transmitted wave.

In the case where more than one wall exists between the transmitter and the receiver, a detailed analysis is required to calculate the total loss (ΣL_w). In this case, the actual geometry of the walls with respect to the incident wave must be considered as well as the construction materials of each wall.

Measurements conducted by [4], at 1700 MHz, in two different types of indoor environments where NLOS exist conclude that the distance-dependent mean path loss is given by:

$$M(d) = n\,10\log(d) + K(dB) \tag{6.4}$$

where d is the distance in meters between the reference point S and the receiver.

For the indoor environment with NLOS due to some thin wall or shadowing objects, $n = 4.1$ and $K = -19$ dB.

For the indoor environment with NLOS and with many small rooms and thick walls $n = 6$ and $K = -36$ dB.

6.2.3 Location: Tx At a Different Floor From the Rx—Both Inside the Same Building

Case 3: B3 to M1 and M2 (Figure 6.2) In case 3, the transmit base and the receive mobile are within the same building but at different floors. In this case the path loss is given by:

$$P_3 = S + 10\,n_3\log(d_1) + K_3 F_3 + 10\,n_3'\log(d/d_1) \tag{6.5}$$

where:

d = the distance, in meters, between base and mobile without any blockage;
d_1 = the distance to the roof;
F_3 = the floor attenuation factor, which depends on the type of materials and the construction of the floor;
k_3 = the floor number between transmitter and receiver;
n_3 = the environmental dependent exponent in the first floor;
n_3' = the environmental dependent exponent of the second floor.

For k floors we will have:

$$P_3 = S + \Sigma F_k + 10\Sigma n_k\log(d_k) \tag{6.6}$$

where n_k is the environmental dependent exponent for the floor k.

E. H. Walker [6] shows that at 1.7 GHz a floor penetration factor $F_3 = 14.8$ dB. The floor consists of hollow pot terra-cotta tiles covered by 15-cm metal mesh and wooden floor tiles. It is pointed out that a signal decay index n_3 of around 3.5 applies in the horizontal plane. In general, once the floor loss has been taken into account there is little to attenuate the signal in the vertical plane. Also, there is strong evidence of signal power being channeled up stairwells and lift shafts at 1.7 GHz.

Measurements are required to determine the values of F_3 and n_3. These measurements must include the type of floors and the effect of signals channeled up stairwells and lift shafts. From the measurements, we will also determine if the value of n_3 is correct for both the horizontal and the vertical plane.

Signal-level measurements were made in two dissimilar office buildings [7] at 850 MHz and 1.7 GHz. The model derived from these measurements was a simple free space plus linear path attenuation. As the signal expands from the transmitting antenna it passes through walls and other obstructions. As the number of these obstructions increases, their effect may be approximated by some attenuation coefficient in dB/m. The fitted model curves showed a smaller linear-path attenuation coefficient as the frequency was increased.

Table 6.1 shows the regression parameters for the free space plus linear-path attenuation model, for the measured locations.

Table 6.1
Free Space Plus Linear-Path Attenuation Model

Location	Frequency	Attenuation (dB/m)	Standard deviation (dB/m)
Large building	850 MHz	0.62	8.4
Office	1.7 GHz	0.57	8.5
Medium-size building	850 MHz	0.48	8.0
Office building	1.7 GHz	0.35	9.5

6.2.4 Location: Tx Outside the Building, Rx Inside the Building

Case 4: B4 (Street Level) and B5 (Above Roof Tops) to M3, M4, and M5 (Figure 6.2) In the case where the transmitter is located outside the building and the receiver within the building, on different floors, the propagation loss is given by:

$$P_4 = S + 10n_0 \log(d_1) + L_w + 10n_4 \log(d/d_1) + kM \qquad (6.7)$$

where:

S = 37 dB for 1.7 GHz;
n_0 = the environmental dependent exponent outside the building;
n_4 = the environmental dependent exponent inside the building;
L_w = penetration loss due to external wall (see Table 6.3) (to include internal walls refer to internal model, case 2);

d_1 = the distance between transmitter and the wall outside the building;
d = the distance between transmitter and receiver;
M = the floor dependent factor (floor height effect) and k is the number of floors (M is a function of k).

The parameter L_w depends on the following conditions:

1. Different types of outside wall construction, such as steel-framed glass, brick, block masonry, and so forth.
2. Urban versus suburban areas to identify the difference between buildings in exposed locations and buildings sheltered in the dense urban core.
3. Different building orientations with respect to the transmitting antenna.
4. Different percentages of window areas in the outside walls.
5. Different types of window treatments used to reflect sunlight and heat.

Some measurements were conducted at 1,700 and 900 MHz into a three-story building divided into offices and laboratories. The results, in general, support the model proposed in [6]:

$$P_{loss} = C + 10n \log r + kF \qquad (6.8)$$

where C is the power loss at 1m, n is the signal decay index, r is horizontal range between the transmitter and receiver, k is the number of floors, and f is the individual floor loss factor.

Signal decay index (n) is approximately 3.5 at both 1,700 and 900 MHz. Floor loss F is 12 dB at 900 MHz, but there is an additional 2.8 dB floor-loss factor at 1,700 MHz. The 1m intercept point is approximately 29 dB at 1,700 MHz and 23 dB at 900 MHz. In general, wall loss at 1,700 MHz was 3–4 dB for double plasterboard and 7–9 dB for breeze block or brick.

One of the models [8] that can be used in order to describe the path loss as a function of the distance between transmitter and receiver is given by:

$$P = S + 10n \log(d) \qquad (6.9)$$

The above equation can be modified as follows:

$$P' = S + 10n \log(d) + kf \qquad (6.10)$$

where

P = path loss (dB);
P' = path loss including the floor loss factor;
S = path loss in 1m (dB);

n = power low index;
d = distance between transmitter and receiver;
k = number of floors traversed;
F = signal attenuation provided by each floor.

Two different kinds of measurements were made [8]: at 900 MHz, $F = 10$, $S = 16$, $n = 4$. At 1700 MHz, $F = 16$, $S = 21$, $n = 3.5$.

The above tests were performed in a modern office. The building was of standard steel-frame construction with brick external walls and plasterboard internal partition.

A portable transmitter was moved around selected rooms throughout the building while a stationary receiver was located at the center of the office block. The receiver was attached to a dipole antenna, whereas the portable transmitter could be operated at 900 MHz and 1,700 MHz using appropriate quarter-wavelength with ground plane antennas.

Measurements conducted by [9] with the transmitter located outside of the building and at height of 6 m conclude that the best fit was achieved by:

$$\text{path loss} = L'(v) + 10\,n'\log(d) + kF' + pW_i' + W_e \tag{6.11}$$

where

$L(v)$ = mean of log normal distribution with variance v (dB);
k = number of floors separating Tx and Rx;
F = floor loss factor;
p = number of internal walls separating Tx and Rx;
W_i = internal wall loss factor ($0.4 < W_i < 8$);
W_e = external wall loss factor ($3.8 < W_e < 10.5$);
$'$ = indication of frequency dependence.

The parameters were calculated by iteratively changing the parameters until the lowest standard deviation of error (measured path loss less path loss predicted by (6.11)) and zero mean error were obtained. The external wall factor was extracted from the mean of the log normal by comparison with measurements taken outside the building. The power law n was preset to two while the other parameters were varied. The parameters in (6.11) are given in Table 6.2. The building description is given in Section 6.4.

Measurements have been conducted by [1] to study the penetration loss of different types of walls at 1.7 GHz. The measurements were made with the transmitter on the outside and the receiver on the inside of six different buildings. The transmitter antenna was mounted on a 10m high mast, which is assumed to be the approximate height of future microcellular base-station antennas. The buildings measured were located in the suburbs of Stockholm and five of them were school buildings.

The measured maximum, minimum, and mean values of the wall(s) penetration factor L_w are shown in Table 6.3. As can be seen, the wall attenuation is mainly affected by the material in the wall and, if present, the size of the windows.

Another important factor is the angle of incidence of the transmitted wave (i.e., the angle between the normal of the wall and the straight line to the transmitter). If the angle is wide, much less of energy seems to penetrate the outer wall. This is probably due to three factors:

1. The signal has to travel a longer distance through the attenuating material.
2. The power density is lower at the wall.
3. The energy is reflected by the wall.

Table 6.2
Parameters Used in Equation (6.11)

Building	Cell Type	Frequency (MHz)	L(v) (dB)	n	F	W_i (dB)	W_e (dB)
B81	Near	1,735	39.8	2	0	3.8	3.9
B81	Far	1,735	54.2	2	0	3.8	3.9
B29	Near	1,735	50.2	2	11	0.7	7.5
B29	Far	1,735	50.7	2	12	0.5	6.3
SVH	Far	1,735	52.3	2	0.4	8.0	10.5
BSH	Near	1,735	59.1	2	1	0.9	5.1
BSH	Far	1,735	63.7	2	0.2	0.9	5.1

Table 6.3
Values of Wall Penetration Factor Lw

Wall Type	L_w (dB)	L_w Minimum (dB)	L_w Maximum (dB)
Thick (25 cm) concrete with large windows	5	4	5
Thick concrete with large windows, large incidence	11	9	12
Thick concrete with no windows	13	10	18
Double (2 × 20 cm) concrete indoor	17	14	20
Thin (10 cm) concrete indoor	6	3	7
Brick wall (25 cm) with small windows	4	3	5
Steel wall (1 cm) with large reinforced windows	10	9	11
Glass wall	2	1	3
Reinforced glass	8	7	9

Measurements were also made by [1] to study the large-scale (concerning mean values) cross-polarization coupling effects. The transmitted signal was vertically polarized and the transmitter was placed 100–300m away from the buildings. At the receiver, both vertical local mean L_v and horizontal local mean L_h were calculated at every point measured.

The cross polarization coupling effect for mean values, XPOL, is defined as:

$$XPOL = L_h - L_v \qquad (6.12)$$

where L_h = the horizontal mean value level and L_v = the vertical mean value level.

As can be seen in Figure 6.3, a high degree of XPOL exists in every building. The XPOL increases with decreasing L_v. This XPOL coupling will reduce the effects of random antenna orientation in regions with low signal strength by equalizing the horizontal and vertical signal levels.

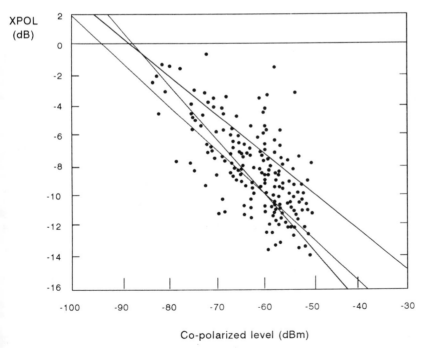

Figure 6.3 Cross-polarization coupling in indoor environment for measured mean values.

Typical construction element attenuation at 900 MHZ is given as [10] and the attenuation is illustrated in Table 6.4.

On higher floors the signal exhibits a height gain relative to the signal at the street level. This height gain can be estimated from Figure 6.4, which is based on the attenuation measurements made in Stockholm, Sweden [11]. The measurements were made in a small-cell environment with a transmit antenna height at 30m and at two frequencies, 955 MHz and 1,705 MHz. Signal-strength levels were recorded outside and inside five different buildings to study the wall penetration loss and at different floor levels to evaluate the height dependence.

Similar results are given in [12] where the path-loss reduction is given by

$$\text{loss (dB)} = 16 - 2 \times \text{(floor level)} \tag{6.13}$$

This path-loss reduction is due to the fact that the Tx antenna is located on top of a building.

Nine experiments were conducted to assess the effect of frequency, transmission condition, and building construction on transmitted into buildings. The measurements were made for three different frequencies (441, 896.5, and 1,400 MHz), whereas the transmitter and receiver locations chosen were within the University of Liverpool.

Measurements conducted by [12] show a floor factor for all floors around of 2.5 dB for a complete line-of-sight, but penetration loss varies for different floors with a partial or no line-of-sight.

Due to the complexity and the number of factors involved in Case 4, a large number of measurements are required for the final values of the parameters in (6.13).

Some measurements were made by the Advanced Mobile Phone Service Developmental Cellular System in Chicago [13] using ten separate set-up channels in which both the receiver and the transmitter could be tuned.

Table 6.4
Typical Construction Element Attenuation

Material	Attenuation (dB)	Standard Deviation (dB)
8-inch concrete wall	7	1
Wood and brick siding	3	0.5
Aluminum siding	2	0.5
Metal walls	12	4
Attenuation past office furnishing (dB/m)	1	0.3

A group of buildings in the Chicago area was selected to present a range of physical characteristics including location, architecture, and local environment. Single strength measurements were made along planned routes on selected floors of these buildings. A propagation measuring set was used to measure the strength of signals transmitted from cell sites. Building penetration loss for a given floor area is defined to be the difference between the average of these measurements and an average of measurements made on the outside at street level.

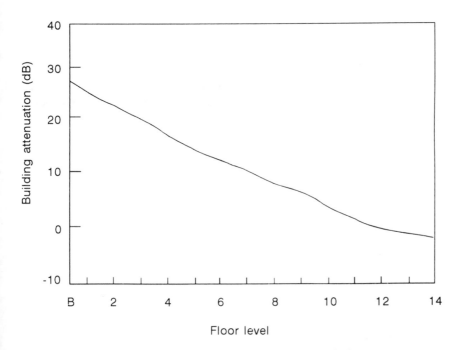

Figure 6.4 Variation with height of the signal reduction inside buildings [11].

Outside signal strength was measured at street level around the perimeter of the building, along the closest available path to the building's outside walls. These paths included driveways streets or parking lots as required to achieve proximity to the building under test.

The given model is:

$$\text{loss (dB)} = 14.2 - 2.6966 * (\text{floor number}) \tag{6.14}$$

The effect of windows is given in [14] where the penetration loss is given as a function of the arrival angle as:

$$L = L_{ow} + 20 \log[1 - G(u)] \qquad (6.15)$$

where L_{ow} is the penetration loss of the received signal arriving from $u = 3$ degrees, and changes according to the situation of the window. These values are 6.6 dB for ordinary state window and 8.9 dB for opening half.

The value of $G(u)$ as a function of arrival angle is shown in Figure 6.5 [14].

6.2.5 Summary of Models

In this section an attempt is made to present a summary of the models analyzed before. To visualize the effect of the parameters involved in the equations of Section 6.2, the models are presented in the form of graphs.

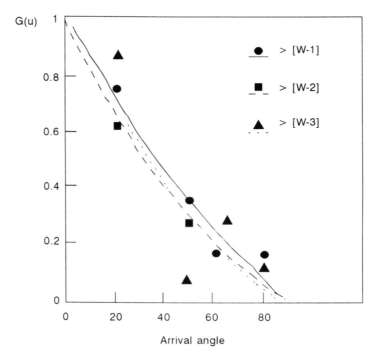

Figure 6.5 Value of $G(u)$ as a function of arrival angle.

From [6] and (6.8):

1,700 MHz	$L(\text{dB}) = 29 + 35 \log(d) + 14.8\,k$	[Model 1]
900 MHz	$L(\text{dB}) = 23 + 35 \log(d) + 12\,k$	[Model 2]

From [8] and (6.10):

1,700 MHz	$L(\text{dB}) = 21 + 35 \log(d) + 16\,k$	[Model 3]
900 MHz	$L(\text{dB}) = 16 + 40 \log(d) + 10\,k$	[Model 4]

From [9] and (6.11):

Building B29

1,735 MHz

NEAR	$L(\text{dB}) = 50.2 + 20 \log(d) + 11\,k + 0.7\,P + 7.5$	[Model 5]
FAR	$L(\text{dB}) = 50.7 + 20 \log(d) + 12\,k + 0.5\,P + 6.3$	[Model 6]

900 MHz

NEAR	$L(\text{dB}) = 39.6 + 20 \log(d) + 8\,k + 0.4\,P + 3.8$	[Model 7]
FAR	$L(\text{dB}) = 38.8 + 20 \log(d) + 8\,k + 0.4\,P + 3.8$	[Model 8]

Building SVH

1,735 MHz

FAR	$L(\text{dB}) = 52.3 + 20 \log(d) + 0.4\,k + 8\,P + 10.5$	[Model 9]

Building BSH

1735 MHz

NEAR	$L(\text{dB}) = 59.1 + 20 \log(d) + 1\,k + 0.9\,P + 5.1$	[Model 10]
FAR	$L(\text{dB}) = 63.7 + 20 \log(d) + 0.2\,k + 0.9\,P + 5.1$	[Model 11]

Building B81

1,735 MHz

NEAR	$L(\text{dB}) = 39.8 + 20 \log(d) + 0\,k + 3.8\,P + 3.9$	[Model 12]
FAR	$L(\text{dB}) = 54.2 + 20 \log(d) + 0\,k + 3.8\,P + 3.9$	[Model 13]

From [13] and (6.14):

$$L(\text{dB}) = 14.2 - 2.6966\,k \qquad \text{[Model 14]}$$

From [12] and (6.13):

$$L(\text{dB}) = 16 - 2\,k \qquad \text{[Model 15]}$$

where

L = penetration loss in dB;
k = number of floors;
P = number of internal walls;
d = distance between Tx and Rx.

6.2.5.1 Floor Factor Effect

Based on (6.8) and (6.10), the effect of floor factor is shown in Figures 6.6, 6.7, and 6.8, as a function of distance (between transmitter and receiver) and frequency.

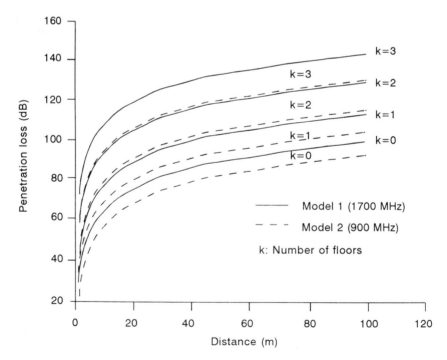

Figure 6.6 The effect of floor loss factor in different frequencies.

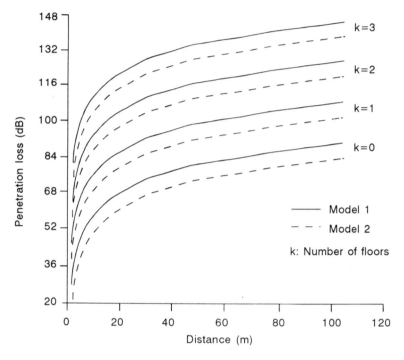

Figure 6.7 The effect of floor loss factor at 1,700 MHz.

In Figure 6.9 we compare two linear models (14 and 15), dependent only on the floor loss factor. The results show negative values of penetration loss in upper floors, which does not mean that we have gain but a decrease of attenuation level compared to the attenuation of the first floor level.

6.2.5.2 Internal Walls Effect

The effect of internal walls, given in (6.11), is presented in Figures 6.10, 6.11, and 6.12 as a function of distance, frequency, and number of walls.

6.2.5.3 Comparison of Models

If we are interested in the similarities and differences among several of earlier presented models we can examine Figures 6.13 through 6.16, where penetration loss is given as a function of distance, floor level varies from 0 to 3, and frequency

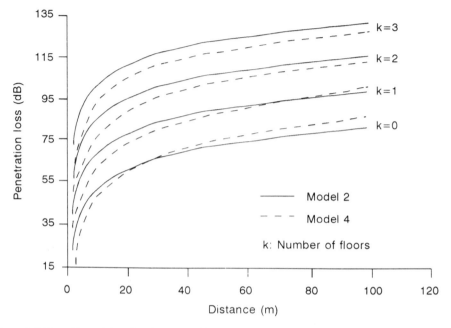

Figure 6.8 The effect of floor loss factor at 900 MHz.

is at 1,700 MHz. For those models in which there is an internal wall loss dependence, we assume that there are three internal walls.

6.3 INDOOR MODEL FOR IMPULSE RESPONSE

6.3.1 Channel Model

The mobile channel is represented by multiple paths having a gain a_k, a propagation delay t_k, and associated phase shift θ_k, where k is the path index. Thus, the complex low pass channel impulse response is given by:

$$h(t) = \Sigma \, a_k e^{j\theta_k} \delta(t - t_k) \tag{6.16}$$

where δ is the Dirac delta function.

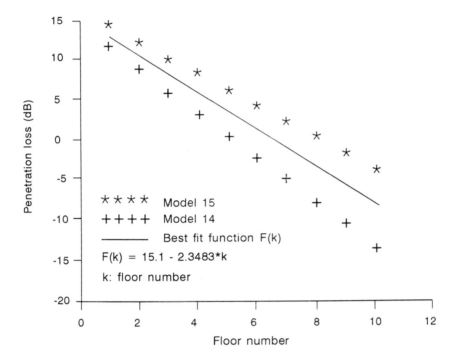

Figure 6.9 Comparison of linear models.

Because of the motion of the people and equipment in and around the building, the parameters a_k, t_k, and θ_k are randomly time-varying functions. However, the rate of their variation is very slow compared to any useful signaling rates that are likely to be considered. Thus, these parameters can be treated as virtually time-invariant random variables.

A convenient model of the impulse response is the transversal filter presentation (Fig. 6.17). The model is completely defined if it is given the gain a_k and the delay t_k of each tap. The parameter θ_k is assumed to be uniformly distributed in $[0-2\pi]$.

6.3.2 Impulse Response Parameters

The error rate for digital transmission through multipath channels is most strongly dependent on the delay spread or width of the power-delay profile. The parameters required to describe some of the important characteristics of the impulse response (IR) of the channel are shown in Figure 6.18 [15].

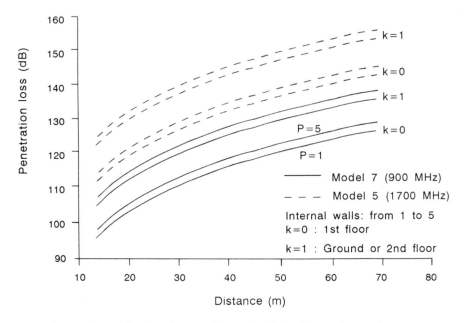

Figure 6.10 The effect of floor loss factor and internal walls in different frequencies.

Mean delay (m) is the power-weighted average of the excess delays, measured and given by the first moment of the IR. Delay spread (s) is the power-weighted standard deviation of the excess delays, and given by the second moment of the IR. It provides a measure of variability of the mean delay. Delay interval (Ix dB) is defined as the length of the IR between two values of excess delay which marks the first time the amplitude of the IR exceeds a given threshold and the last time it falls below it. Delay window ($W\%$ x) is the length of the middle potion of the power profile containing a certain percentage of the total energy found in the IR. NST is effective noise/spurious threshold (noise level + 3 dB). PSR is peak-to-spurious ratio $\geq$ 15 dB. L is legitimate impulse response length. $I(x$ dB) is delay interval for a threshold of x dB below peak. $W(y\%)$ is $y\%$ delay window. E contains $(100\text{-}y)/2$ % of the total energy found in L.

The mean delay and delay spread can be calculated over L as follows:

$$\text{mean delay } (m) = \Sigma\, t_i P_i / \Sigma P_i \tag{6.17}$$

$$\text{delay spread } (s) = [\Sigma(t_i - m)^2 P_i / \Sigma P_i]^{1/2} \tag{6.18}$$

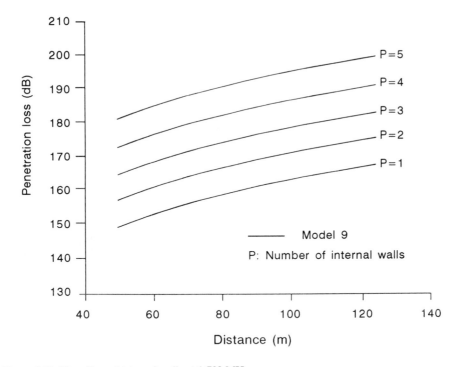

Figure 6.11 The effect of internal walls at 1,700 MHz.

6.3.3 Measurements Locations

To characterize the impulse response of a wideband channel in cells of area less than 0.2 Km², measurements were made [17] in a shopping center, a railway station, and an office block. To obtain information on distance dependence and the effect of LOS, the measured areas were split into "near," "mid," and "far" ranges. Where possible, measurements were made both when an area was full of people, so that their effect could also be examined, and when it was empty.

The transmitter was always used as a fixed station and the receiver as the mobile. The transmitter was always located in a position that was expected to give good coverage of the measurement area. No distance information was possible, so the operator attempted to maintain a constant walking speed of approximately 1.2 m/s within a number of predefined measurement areas. Measures were made in a fixed time (5 minutes, representing over 1,000 impulses responses) during a random walk of the measurement partitioned area.

A detailed description of the measurement locations is given in Section 6.4.

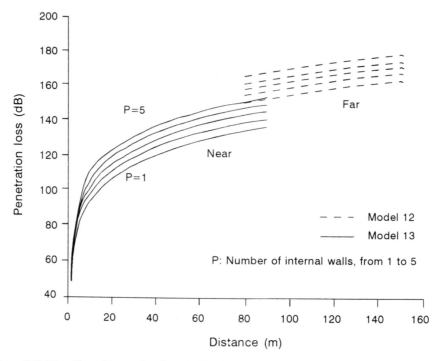

Figure 6.12 The effect of internal walls at 1,700 MHz.

A summary of the results is presented in Table 6.5.

It should be noted that the office block results show larger variations than the other two locations. This is because the radio environment at the shopping center and at the railway station is more homogeneous than environment at the office block [17].

As explained in [17], treatment of the more extreme results requires some care. Areas of very poor signal strength yielding large dispersion parameters should not be given undue emphasis in the analysis. Such areas would probably require a different base-station arrangement to improve the signal-strength coverage, thereby altering the multipath geometry and the dispersion characteristics. This shows the importance of base-station engineering techniques and that dispersion can be limited by confining coverage to a few rooms rather than an entire floor in a wireless PBX application.

Conversely, in areas of poor signal strength, if a prospective radio system fails to receive adequate signal strength to enable its equalizer to work, the degree of dispersion hardly matters.

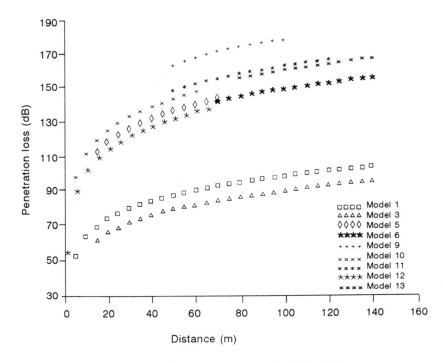

Figure 6.13 Comparison of several models for zero-floor level at 1,700 MHz.

Based upon the 90% confidence limit of the measurements conducted by [17], it is recommended that the following mean delay values, derived from Table 6.5 be used:

- Shopping centers of similar dimensions and construction to the one measured by [17]: mean delay = 150 ns.
- Railway stations of similar dimensions and construction to the one measured by [17]: mean delay = 250 ns.
- Office block of similar dimensions and construction to the one measured by [17]: mean delay = 225 ns (in rare cases rising to 500 ns).

6.3.4 Transversal Filter Representation: Tap Setting

Based on the results of [17], the model for indoor environments is based upon a single-cluster exponential profile, implemented in a linear transversal filter representation of the channel. Tap weights were given for a normalized mean delay (see Table 6.6) with unequal spacing for both six and twelve tap realizations. The

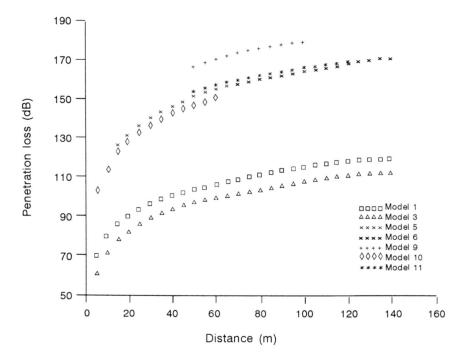

Figure 6.14 Comparison of several models for first-floor level at 1,700 MHz.

statistics of the fast fading along the profile indicates that the first tap should be Rician, with all others being Rayleigh.

6.3.4.1 Typical Impulse Responses

Typical impulses responses and cumulative distributions generated from the measurements of [17] are shown in Figures 6.19, 6.20, and 6.21.

Measurements conducted [18] in an eight-floor office building in New York at 1,700 and 900 MHz show that the delay spread did not exceed 100 ns at either frequency. Because of the layout of the building, most data was line-of-sight path transmitter and receiver. The received power was also normalized by the average power received at a distance 0.3m from the transmitter to give a measure of the relative path loss. Received power levels were also equal at the two frequencies.

The results of measurements in a medium-size two-story building [19] indicate separate clusters of arriving rays covering a 200-ns time span and a delay spread

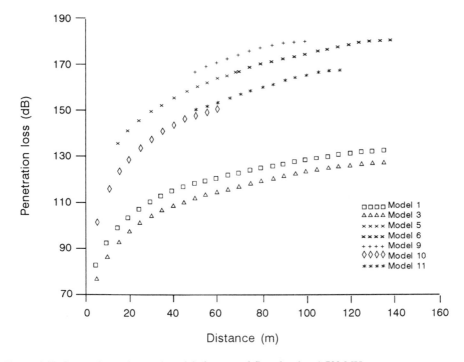

Figure 6.15 Comparison of several models for second-floor level at 1,700 MHz.

up to about 50 ns. The path attenuation follows the distance-power law with exponential decay $n = 3$ which seems to fit room measurements reasonably well.

To characterize the indoor impulse response and classify the different environments where the measurements were conducted requires a detailed description of the indoor environment. General classification, such as large room, typical office, or medium-size hall makes it very difficult, if not impossible, to compare the results of measurements with published results.

6.4 ANNEXES: BUILDING DESCRIPTION

6.4.1 British Telecom

B81 is a three-story L-shaped building. The exterior walls consist of glass with a paint applied, within an aluminum frame and the building is built around a steel framework. The exterior walls contain Astrawall (which is molded metal cladding). The Astrawall in this case is backed by fiberglass insulation. The first two floors

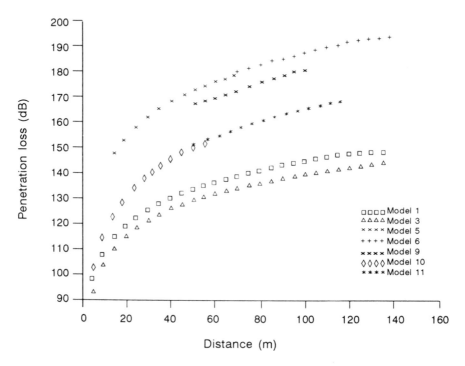

Figure 6.16 Comparison of several models for third-floor level at 1,700 MHz.

are mainly open plan with no partitions above head height. The few rooms that exist are divided by metal partitions. The third floor contains enclosed rooms and laboratories. The building is relatively clear of surrounding clutter due to a car park on one side, a relatively new building under construction on the second side, and a grassy area on the third side.

B29 is a two-story building and has exterior walls made of glass with metal spattering. The building is built around a steel framework. Inside, there are mainly offices and laboratories on the ground floor, but on the first floor there is a mixture of partitions. Only the common corridors could be used for testing. At the top of the building, above the first floor, there is a service area. Only the first two floors were tested. The building is surrounded on one side by steerable antennas and on all other sides by three or four-story buildings.

SVH is a ten-story building. The exterior is a brick tile with cement reinforcement. There is also steel reinforcement in the building. The interior is composed of metal-stud partitioning. Tests were conducted in the common corridors and the lift areas of the building.

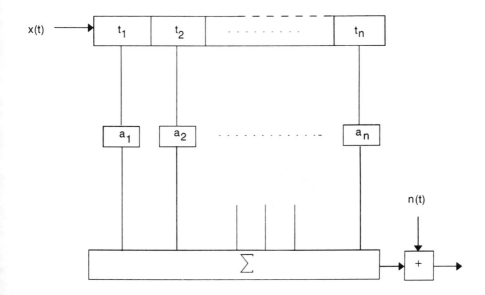

Figure 6.17 Transversal filter presentation.

BSH is an eight-story reinforced concrete building. The inside walls consist of wood and plaster. Only the common areas could be used to conduct measurements in this building.

6.4.2 Shopping Center: Tower Ramparts, Ipswich

Tower Ramparts shopping center in Ipswich is a three-story building of steel-frame construction, with glass-clad frontage and roof. Internal walls are of blockwork and reinforced concrete. The ground floor and first floor are shopping malls, while the third is a restaurant. The shopping malls are approximately 96m long. They funnel from 13m wide to 9m, and then up to 17m in a square at the far end.

The transmitter was located at the top of the stairs leading to the second-floor restaurant at one end of the building. The antenna was then 9.5m above ground floor level. Impulse responses were taken on both ground and first-floor malls. The received antenna was 1.9m high.

6.4.3 Railway Station: Euston Station, London

Euston railway station in London consists of a large open area 34m by 60m and 10m high, with a glass frontage and platforms at the rear. There is a mall at the

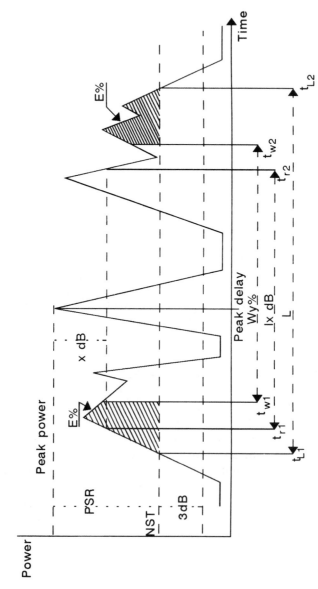

Figure 6.18 Path profile parameters [16]. See text for definitions of parameters.

Table 6.5
Measurement Results From Several Locations

Location	Mean Spread (m, ns)	Delay Spread (s, ns)	50 % Window (W, ns)	6 dB Interval (I, ns)
Shopping center	120–153	115–200	125–164	140–200
Railway station	200–241	200–464	206–309	215–370
Office block within building	107–444	120–464	98–508	100–700
Office block inside to outside	110–200	100–200	84–219	95–240

Table 6.6
Tap Setting

Tap Number	Normalized Delay	Power (dB) (12 Taps)	Power (dB) (6 Taps)	Spectrum Type
1	0	−1.5	−1.5	Rice, $k = 2$
2	0.3	0	0	Rayleigh
3	0.633	−2		Rayleigh
4	1	−3	−2	Rayleigh
5	1.467	−4	−3	Rayleigh
6	1.967	−5.5		Rayleigh
7	2.467	−7		Rayleigh
8	2.967	−8.5	−4	Rayleigh
9	3.533	−10		Rayleigh
10	4.133	−12		Rayleigh
11	4.867	−15	−10	Rayleigh
12	5.533	−20.5		Rayleigh

platform end of the building which is 160m by 8.5m. It is of reinforced concrete construction. The transmitter was used as the fixed station and was located on a balcony in the middle of the glass frontage. The antenna height was 7.5m. The receiver was carried in a rucksack, with the antenna at a height of 1.8m high, at a walking speed of approximately 1.2 m/s. Three median profiles averaged over 500 impulses each were produced, representing near, middle (LOS), and far (NLOS) cases.

6.4.4 Office Block

The research labs consisted of a number of buildings. Two sets of measurements were taken, one contained solely in the fourth floor of the main-lab block, and from the main-lab block to a garage. The main-lab block was of reinforced concrete construction, but had unusual internal-wall partitions going from one room to

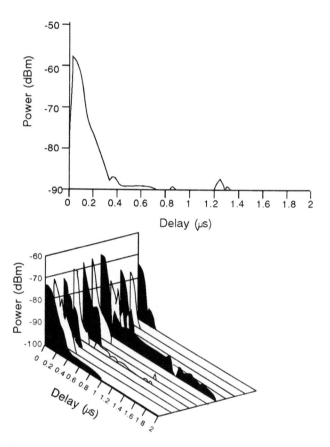

Figure 6.19 Impulse response, shopping center.

another. The transmitter was located by the window in a room at the corner of the floor. The windows of each room represented 50% of the outside-wall surface. They were double-glazed, air-filled, and uncoated.

The transmit antenna height was 1m above floor level and 20m above ground floor level. The receiving antenna was 2m high. Measurements were made in four corridor sections and seven rooms of the fourth floor. All rooms were NLOS, with dimensions approximately 68 m^2. Ceilings were 3.3m above the floor level and were again steel panels. There were approximately six people per room (0.09 per m^2, mainly seated) and, on average, none in the corridors. During the measurements all doors were closed.

The second set of measurements were made in and around the garage. The garage was 20m below the fourth floor. The receiver antenna height was 2m. The

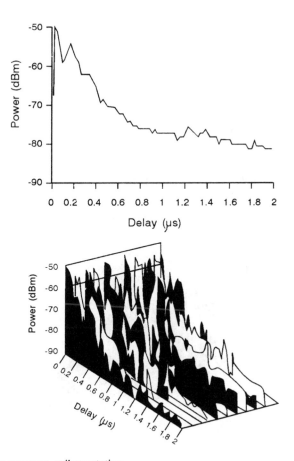

Figure 6.20 Impulse response, railway station.

garage was of prefabricated concrete-slab construction with the building's front of traditional brick. Measurements were made around the garage where LOS and NLOS paths existed. Measurements were then made in the garage with the door both closed and open.

6.5 CONCLUSIONS

The mobile-radio channel for indoor or outdoor environment is randomly dispersive in both time and frequency. The transmitted waves are scattered, diffracted, and attenuated. The various scattered components interfere, building up an irregular

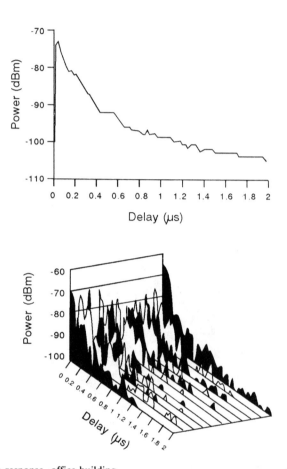

Figure 6.21 Impulse response, office building.

field distribution. The signal at the receiver is therefore attenuated and distorted. The severity of the disturbance depends on the specific physical properties of the scattering environment.

To characterize the mobile channel requires a complete knowledge of the propagation parameters for all environments where the system will operate. Conducting measurements to obtain all propagation parameters for all possible environments is an impossible task and, for a limited number of environments, is a time-consuming exercise.

Therefore, propagation models are used to predict the performance of a mobile-communication system. These models may be either theoretical or empirical, or a combination of the two. Empirical models are derived from measurements and observation and offer a major advantage in that all environmental influences are implicit in the result.

Present and future mobile-communication systems require the development of a channel model(s) for both indoor and outdoor environments. The parameters of these models are:

1. Signal attenuation-path loss.
2. Signal fading (fast and slow fading).
3. Fading depth and rate of occurrence.
4. Impulse response:
   a. Mean delay,
   b. Delay spread,
   c. Delay interval, and
   d. Delay window.
5. Noise environment.
6. Interference environment.

REFERENCES

[1] Bachman, P. O., S. Lidbrink, and T. Ljunggren. "Building Penetration Loss Measurements at 1.7 GHz in Micro Cellular Radio Environments," *RMTP/RB/G006*.

[2] Cox, D. C., R. R. Murray, and A. W. Norris. "800-MHz Attenuation Measured in and Around Suburban Houses," *AT&T B.L.T.J.*, Vol. 63, No. 6, July–August 1984.

[3] Bergljung, C., et al. "Micro-cell Radio Channel," *Preliminary report on Indoor Field-strength. Measurements at 900 and 1700 MHz.*, RMTP/RB/J031.

[4] *Investigation of Radio Propagation and Microscopic Diversity in Indoor Microcells at 1700 MHz.*, RMTP/RB/J082, Issue 1, September 25, 1990.

[5] Lafortune, J. F., and M. Lecours. "Measurements and Modeling of Propagation Losses in a Building at 900 MHz," *IEEE Trans. on Veh. Tech.*, Vol. 39, No. 2, May 1990.

[6] Owen, F. C., and C. D. Pudney. "Radio Propagation for Digital Cordless Telephones at 1700 MHz and 900 MHz," *Electronic Letters*, Vol. 25, No. 1, January 5th, 1989.

[7] Devasirvatham, D. M., et al. "Multi-Frequency Radiowave Propagation Measurements in the Portable Radio Environment," *ICC 90*.

[8] Motley, A. J., and J.M.P. Keenan. "Personal Communications Radio Coverage in Buildings at 900 MHz and 1700 MHz," *Electronics Letters*, Vol. 24, No. 12, June 9th, 1988.

[9] Johnson, I. T., W. Johnston, and F. J. Kelly. "CW and Digital Propagation Behavior from Outdoor Public Cells Measured in Customer Premises Cells," *RMTP/CC/E124*, Issue 1, CEC Deliverable, No. 43/BTR/CCV2/DA/A/058/b1.

[10] IEEE VT Society Committee on Radio Propagation. "Coverage Prediction for Mobile Radio Systems Operating in the 800/900 MHz Frequency Range," *IEEE Trans. on Veh. Tech.*, Vol. 37, No. 1, February 1988.

[11] "Building Penetration Loss Measurements in the Small Cell Environment," *RTMP/RB/G008*, Issue 1, September 20, 1990.

[12] Turkmany, N.M.D., J. D. Parssons, and D. G. Lewis. "Radio Propagation into Buildings at 441, 900 and 1400 MHz,"

[13] Walker, E. H. "Penetration of Radio Signals into Buildings in the Cellular Radio Environment," *B.S.T.J.*, No. 9, November 1983.

[14] Hirokoshi, J., K. Tanaka, and T. Morinaga. "1.2 GHz Band Wave Propagation Measurements in Concrete Buildings for Indoor Radio Communications," *IEEE Trans. on Veh. Tech.*, Vol. VT-35, No. 4, November 1986.

[15] "Measurement, Analysis and Presentation of Propagation Data," *RMTP/RB/E0xx*.

[16] Constantinou, P. "RMTP Propagation Models," *RACE MOBILE 1043, RMTP/RB/S303*, Final, Cambridge, November 1990.

[17] "RB2 Measurements of Wideband Channel Characteristics in Cells within Man Made Structures of Area Less Than 0.2 Km²," *RMTP/RB/E094*, Final Report, November 22, 1989.

[18] Devasirvatham, D.M.J., and R. R. Murray. "Time Delay Spread Measurements at 850 MHz and 1.7 GHz Inside a Metropolitan Office Building," *Electronic Letters*, Vol. 25, No. 3, February 2, 1989.

[19] Constantinou, P. "Mobile Communication—Channel Modelling," *British Telecom Eng.*, Vol. 9, August 1990.

[20] Saleh, A.A.M., and R. A. Valenzuela. "A Statistical Model for Indoor Multipath Propagation," *IEEE Journal on Selected Areas in Communication,* Vol. SAC-5, No. 2, February 1987.

[21] Paunovic, D. S., Z. D. Stojanovic, and I. S. Stojanovic. "Choice of a Suitable Method for Prediction of the Field Strength in Planning Land Mobile Radio Systems," *IEEE Trans. on Veh. Tech.*, Vol. VT-33, No. 3, August 1984.

[22] Gudmunsdson, B. "Microcellular Base Station Topology," *MTP/CM/J039*.

[23] Allesebrook, K., and D. Parsons. "Mobile Radio Propagation in British Cities at Frequencies in the VHF and UHF Bands," *IEEE Trans. Veh. Tech.*, Vol. VT-26, No. 4, November 1977.

[24] Kozono, S., and K. Watanabe. "Influence of Environmental Buildings on UHF Land Mobile Radio Propagation," *IEEE Trans. on Selected Areas of Communication*, Vol. COM-25, No. 10, October 1977

[25] "Characterisation of the mobile channel at 1900 MHz by power delay profiles. First measurements," *RMTP/RB/H404*, Issue 1, July 25, 1989.

Chapter 7

Antenna Systems For Wireless LANs

Christos N. Capsalis

National Technical University of Athens
Dept. of Electrical Engineering, Electroscience Division
Patission 42, Ave. 10682 Athens, Greece

7.1 INTRODUCTION

The provision of data communications to people when they are away from a certain point of attachment has become a major communications frontier. In a business environment especially, the development of wireless computer networks forms a new and interesting problem in the area of mobile communications as the wireless in-house communication data networks are technically quite different from high-power vehicular cellular mobile radio that in many ways use higher power in comparison with in-house terminals. Some of the ways they are different include the frequency to be used, the propagation characteristics, the limited available space, the influence of human body on the performance of the antenna system, and finally the very low radiated-power requirement for safety reasons.

In general, the characteristics of the wireless LANs systems are comparable to those of portable radio systems. Many of these characteristics, listed below, are necessary for choosing the network-terminal antenna system, while others are presented just give a general view of the systems under study.

1. The probability of no coverage within a service area inside a building is less than 10^{-4}.
2. Average portable transmitter power is 5–10 mW.
3. Radio access techniques include: TDMA, ATM, and others, depending on the specific application.
4. There are 5 to 20 sets of channels depending on the specific application.
5. Diversity to mitigate small-scale signal variation (optional).
6. Flexible user bit rates in 8 Kbps increments.
7. Radio-channel transmission rate ranges from 400 Kbps to 10 Mbps gross.

8. The operating frequency, depending, with other factors, on the specific application is: VHF (450 MHz); UHF (900 MHz); 1,800–2,300 MHz; 2,400 MHz; 5,725 MHz; and 18 GHz.

The operating frequencies have been allocated by the FCC for narrow-band communications inside factories, offices, and so forth, while the frequencies 1,800–2,300 MHz are dedicated for universal mobile telecommunication systems (UMTS). On the other hand, Motorola's wireless in-building network uses low-power microwave at 18 GHz. Recently the potential use of these frequencies for wideband radio LANs has been recognized. A recent announcement has also drawn attention to the 28-GHz band for curb-to-home high-speed channels. For broadband in-house propagation, the operating frequency of 60 GHz is also under study.

In general, both the network and terminal antenna systems may be similar. This is in fact only a theoretical statement because at the terminal there are severe limitations of volume and weight, so only small antennas are used at the terminal side. On the other hand, in most cases an almost-omnidirectional antenna may be used at the terminal while at the network side, in many cases, a directional antenna system may be used.

In this chapter antenna systems that may be used efficiently in wireless LAN applications are described. Special attention is given to the terminal antenna in order to overcome the multipath propagation in the in-house environment.

Simple antennas are described first because they can be used either individually or in combination (arrays, diversity antenna systems) to form an efficient transmitting/receiving system for wireless LANs applications.

Also, the principles of directional antennas are presented because depending on the coverage area (e.g., long corridors), in many cases in-house wireless communications antenna systems with considerable directivity may be used. Depending on the degree of mobility, directional antennas can also be used effectively at the mobile terminal, especially for applications where the user has low mobility when using the equipment (i.e., wireless data terminals).

Printed antennas are briefly described as they can be used in both the network and the terminal because they offer many advantages, for example, low weight and volume, low cost, they can be easily mounted on the mobile terminal, they have low scattering cross section, the changing of the polarization (linear, circular) is quite easy with simple changes in feed position, and they are compatible with modular designs when feed lines and matching networks are fabricated simultaneously with the antenna structures.

Special attention is given to the examination of using microscopic and/or macroscopic diversity techniques in order to overcome the indoor multipath fading. After presenting these concepts, some sophisticated antenna systems based on the diversity principle are briefly described.

In the final part of this chapter the influence of the human body on the antenna's performance is examined through the variation of the antenna's reflection coefficient (S_{11}) when the antenna system is inside the real office environment. Finally, the statistics of the variation of S_{11} are discussed.

7.2 SIMPLE ANTENNAS

7.2.1 Wire Antennas

The most common antennas used in portable communication systems are the wire antennas, namely dipoles of various lengths because they are simple and low-cost antennas. Also, the dipoles can be used as basic structural elements to form arrays or space diversity schemes. In addition, loop antennas can be used in terminal equipment because they can be (possibly) used in flat structures. In the rest of this paragraph the basic properties of the small dipole and the dipole with arbitrary length are briefly described and the major antenna measures are defined.

In the study of antenna systems reference is usually made to the simplest dipole, which is commonly known as "small dipole" or "Hertzian dipole." This antenna is simply a very short length of conductor so that we can suppose that the current distribution is constant throughout its length. In Figure 7.1 the small dipole antenna is shown at the origin of spherical coordinates.

Following well-known procedures [1–3] the far-field radiation field distribution has the following form:

$$E_v = 2jkI_o l \cdot \sin v \cdot e^{-jkr}/(4\pi r) \tag{7.1}$$

$$H_\phi = E_v/\eta \tag{7.2}$$

where $\eta = (\mu/\epsilon) = 120\pi$ ohms for free space, E_v, H_ϕ represent the electric and magnetic fields, $k = 2\pi/\lambda$ is the wave number, and $j = \sqrt{-1}$. As can be observed from (7.1), the radiation pattern of this elementary dipole has a maximum at $v = \pi/2$ and diminishes at $v = 0$ or $v = \pi$. Also, this simple antenna has omnidirectional radiation with respect to ϕ.

The Poynting vector, which is defined in general as $\overline{P} = \overline{E} \times \overline{H}$, is completely in the radial direction, and the time average is given as the relation,

$$P_r = \eta k^2 I_o^2 (2l)^2 \sin^2 v/(32 \cdot \pi^2 r^2) \ W/m \tag{7.3}$$

while the total energy flow can be computed easily as follows:

$$W_{av} = 40 \cdot \pi^2 I_o^2 (2l/\lambda)^2 \ W \tag{7.4}$$

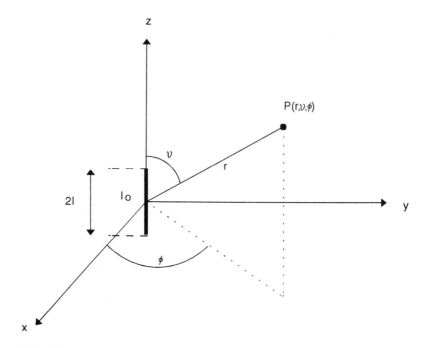

Figure 7.1 Small dipole antenna.

The radiation resistance, which is defined as the resistance, which would dissipate the same amount of power with the same constant current flowing, is given by the relation

$$R_r = 80 \cdot \pi^2 (2l/\lambda)^2 \ \Omega \qquad (7.5)$$

Note that the polarization of any dipole is linear (i.e., the same as the direction of its axis).

All antennae exhibit directive effects, that is, the intensity of radiation is not the same in all directions from the antenna. The property of radiating more strongly in some directions is called directivity and is defined as the ratio of the maximum power density to the average density taken over the entire sphere,

$$D = P/P_{av} \qquad (7.6)$$

In the case of the small dipole antenna the directivity is easily calculated and is $D = 1.5$. Closely related to the directivity of an antenna is another antenna characteristic, the "antenna's gain." This measure takes into account, except from the directive properties, the power losses that may occur in the actual antenna system,

$$G = n \cdot D \qquad (7.7)$$

where n is the efficiency of the antenna system, namely the radiated power divided by the power input.

In cases where the antenna length is not negligible in comparison with the wavelength, the antenna's properties are quite different from the small dipoles. In this case the current distribution of the dipole is not any more constant and a well-known assumption for its distribution is [1–3]:

$$I(z) = \begin{cases} I_m \sin[k(l - z)] & z > 0 \\ I_m \sin[k(l + z)] & z < 0 \end{cases} \qquad (7.8)$$

In this case, the field distribution is given as follows,

$$E_v = j\eta I_m e^{-jkr}[\cos(kl \cos v) - \cos kl]/[2\pi r \sin v] \qquad (7.9)$$
$$H_\phi = E_v/\eta$$

The time average of the Poynting vector can be computed using the relation,

$$P_r = |E_v||H_\phi|/2 \qquad (7.10)$$

while the total radiated power and the radiation resistance are given as follows:

$$W_{av} = nI_m^2/4\pi \int_0^{2\pi} [\cos(kl \cos v) - \cos kl]^2/\sin v \, dv \qquad (7.11)$$

$$R_r = 2W_{av}/I_m^2 \qquad (7.12)$$

In the case of a $\lambda/2$ dipole, the radiation pattern has a maximum at $v = \pi/2$ while it becomes zero at $v = 0$ or $v = \pi$. Also the directivity of the $\lambda/2$ dipole is $D = 1.64$ and the radiation resistance becomes $R_r = 73.09$ Ohms. For longer dipoles, the radiation pattern may comprise one or more side lobes except those of the main lobe, in addition to the main lobe.

A characteristic that plays a major role in the design of any antenna system is the input impedance, which in general consists of the antenna's self-impedance, that is, the impedance that would be measured at the input terminals of the antenna in free space and the mutual impedance that arises from the coupling, either between other antenna elements (e.g., Yagi configuration) or between the antenna and reflecting obstacles.

On the other hand, the self impedance has both resistive and reactive components and, finally, the resistive component comprises the radiation resistance and the loss resistance, which accounts for the dissipative and ohmic losses in the antenna structure.

Only the radiation resistance of any dipole is given in (7.12). Also, the reactance of any dipole can be computed approximately and results are presented in text books [1–4]. In practice, the total input impedance can be measured in terms of the antenna's reflection coefficient, r, and can be expressed as follows,

$$Z_{in} = Z_0(1 + \rho)/(1 - \rho) \tag{7.13}$$

where Z_0 represents the characteristic impedance of the transmission line that is connected to the antenna system. In the final section of this chapter the antenna's reflection coefficient is used as a parameter in order to study the influence of the office environment on the performance of the antenna.

7.2.2 Slot Antennas

According to the basic theory of electromagnetism, fields across an aperture may excite radiation in space. When the apertures are small in order to radiate appreciable amounts of power, they must generally be resonant. A narrow resonant halfwave slot has many similarities to the halfwave dipole, but the electric and magnetic fields are interchanged. In the literature the reciprocity between results for the slot antenna and the dipole follow from the well-known "Babinet's principle" that is an extension of the principle of duality.

The slot antenna may prove to be an efficient radiator for the wireless LANs as it requires minimum available space (of course depending on the frequency of operation, but above 900 MHz its dimensions are quite convenient for use) and also it can be used in combination with a microstrip antenna to form a field diversity scheme.

In order to study the characteristics of the slot antennas, a rather simple example (Fig. 7.2) is examined. The radiation pattern of this antenna system may be determined from the currents running on the conducting surface, but in general a slot in an infinite-ground plane is equivalent to a distribution of magnetic currents confined to the slot area and the radiated fields may be determined by integration over the slot area. In the case of the center-fed slot configuration shown in Figure 7.2, the far electric-field components can be expressed as follows [1–4]:

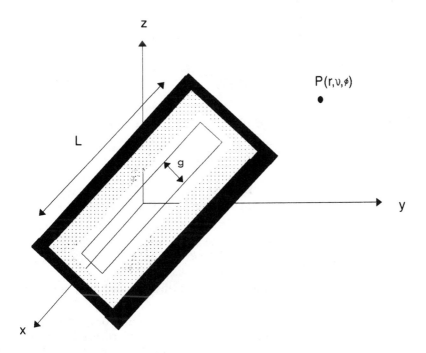

Figure 7.2 Narrow-slot antenna.

$$E_v = (-jk_o/4\pi)[\exp(-jk_o r)/r] \int_{-g/2}^{g/2} \int_{-L/2}^{L/2} \{(-M_x \sin \phi)$$
$$\cdot \exp[jk_o(x \cdot \sin v \cdot \cos \phi + y \cdot \sin v \cdot \sin \phi)]\} dx \, dy \quad (7.14)$$

$$E_\phi = (jk_o/4\pi)[\exp(-jk_o r)/r] \cos v \int_{-g/2}^{g/2} \int_{-L/2}^{L/2} \{(M_x \cos \phi)$$
$$\cdot \exp[jk_o(x \cdot \sin v \cdot \cos \phi + y \cdot \sin v \cdot \cos \phi)]\} dx \, dy \quad (7.15)$$

where M_x represents the magnetic surface current that can be expressed in terms of the slot electric field E_y. Note that E_y is the only component of the slot electric field distribution. If the slot dimensions are supposed to be much less than the wavelength, we may assume that the slot electric field E_y has constant value (i.e., $E_y = E_0$). In this case, $M_x = E_0$ and (7.14) and (7.15) reduce to the following form:

$$E_v = \frac{jk_o ELg}{4\pi} \frac{e^{-jk_o r}}{r} \frac{\sin[(k_0 L/2) \cdot \sin v \cdot \cos \phi]}{(k_0 L/2)\sin v \cdot \cos \phi} \cdot \sin \phi \qquad (7.16)$$

$$E_\phi = \frac{jk_0 ELg}{4\pi} \frac{e^{-jk_o r}}{r} \frac{\sin[(k_0 L/2) \cdot \sin v \cdot \cos \phi]}{(k_0 L/2) \cdot \sin v \cdot \cos \phi} \cdot \cos \phi \cdot \cos v \qquad (7.17)$$

Then the radiation pattern can be obtained using the relation,

$$U_E(v, \phi) = |E_v|^2 + |E_\phi|^2 \qquad (7.18)$$

Note that the radiation pattern diminishes for $v = \pi/2$ and $\phi = 0$.

The radiation conductance may be interpreted in terms of the maximum gap voltage (gE_0) as follows:

$$G_{slot} = 2W/(g \cdot E_o)^2 \qquad (7.19)$$

where W represents the total radiated power that results after integrating the Poynting function.

Also, the radiation resistance of the slot antenna has been evaluated in [15] in the practical case that the slot is excited by an open microstrip line with current distribution $I(x) = |x|$ and the results are presented elsewhere [11,15]. For further discussion on practical slot antenna systems (construction, material, shape, etc.) the reader is referred to the literature [3,11,14,15,16].

7.2.3 Microstrip Antennas

Various types of flat-profile printed antennas can be used for wireless in-house communication systems. In this section the microstrip antennas are briefly discussed, namely the microstrip antennas and the printed dipole antennas. The following characteristics can be identified [3,11,14]:

- The profile of the microstrip and printed dipole antenna is thin.
- The fabrication procedures are very easy (e.g., photolithography).
- The polarization can be either linear or circular. For most applications, linear polarization is adequate. However for satellite communications and some other applications, circular polarization may be necessary. In this case the excitation of two orthogonal modes in the patch elements should be realized in a simple form to avoid complicated feeds.
- Dual frequency operation is possible only for the microstrip antennas, which offer the best shape flexibility as they can be constructed in any shape.
- On the other hand, spurious radiation exists.

- The bandwidth of the above-described antenna types is around 15 MHz when the operating frequency is 900 MHz and it can be broadened when the operating frequency becomes higher.

The following advantages can be identified:

- Low weight and volume.
- Low cost.
- Easy to mount on the mobile terminal.
- Low scattering cross section.
- Changing the polarization (linear, circular) is quite easy with simple changes in feed position for the microstrip antennas.
- Microstrip antennas are compatible with modular designs (i.e., solid-state devices can be added directly to the antenna substrate board).
- Feed lines and matching networks are fabricated simultaneously with the antenna structures.

Some disadvantages of microstrip antennas can be identified and need to be taken into account when selecting the most appropriate antenna type for both the network and the terminal.

- The microstrip antennas have narrow bandwidth in comparison to conventional microwave antennas.
- Higher loss.
- Half-plane radiation.
- Limited maximum gain <18 dB.
- Poor end-fire radiation performance.
- Low power-handling capability and
- Poor isolation between the feed and the radiating elements.

The simpler microstrip antenna configuration is the rectangular patch antenna (which is used by Suite 12 Group at 28 GHz). See Figure 7.3. Its radiation properties are presented as an example in the rest of this paragraph. A lot of models have been developed to study the properties of the microstrip antennas. The results that are presented here come from the so-called empirical models. For an exhaustive discussion on this subject the reader is referred to [3,11,14,17].

The electric far field is given by the relations

$$E_v = K \frac{e^{-jk_o r}}{r} \cdot \cos(k_o h \cdot \sqrt{\epsilon_r} \cos v) \cos \phi$$

$$\cdot \frac{\sin[(\pi \cdot W/\lambda_o)\sin v \cdot \sin \phi] \cdot \cos[(\pi L/\lambda_o)\sin v \cdot \cos \phi]}{\sin v \cdot \sin \phi} \quad (7.20)$$

and

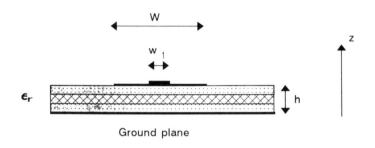

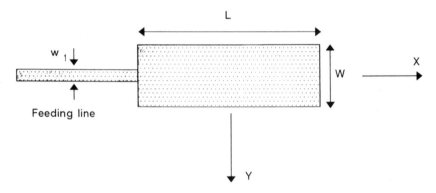

Figure 7.3 End-fed rectangular patch antenna.

$$E_\phi = K \frac{e^{-jk_o r}}{r} \cdot \cos(k_o h \cdot \sqrt{\epsilon_r} \cdot \cos v) \cdot \cos v$$

$$\cdot \frac{\sin[(\pi \cdot W/\lambda_o) \cdot \sin v \cdot \sin \phi] \cdot \cos[(\pi \cdot L/\lambda_o) \cdot \sin v \cdot \cos \phi]}{\sin v} \quad (7.21)$$

while the magnetic field distribution can be easily derived from the electric field distribution.

The directivity of this antenna type can be computed by the relation,

$$D = 8 \cdot W^2 \pi^2 / (I_1 \lambda_0^2) \quad (7.22)$$

where

$$I_1 = \int_0^\Pi \sin^2(\pi \cdot W \cdot \cos v/\lambda_0) \cdot \tan^2 v \cdot \sin v \cdot dv \qquad (7.23)$$

The radiation resistance is

$$R_r = 120 \cdot \pi^2 / I_1^2 \qquad (7.24)$$

and, finally, the resonance frequency of the structure under consideration is

$$f_r = (c/2)\sqrt{\epsilon_{re}}(L + \Delta l) \qquad (7.25)$$

where,

$$\epsilon_{re} = \frac{\epsilon_r + 1}{2} + \frac{\epsilon_r - 1}{2} \cdot (1 + 10h/W) \qquad (7.26)$$

and

$$\Delta l = 0.824h \frac{(\epsilon_{re} + 0.3)(W/h + 0.264)}{(\epsilon_{re} - 0.258)(W/h + 0.813)} \qquad (7.27)$$

7.3 DIRECTIONAL ANTENNAS

Depending on the coverage area (e.g., long corridors), in many cases in-house wireless communications antenna systems with considerable directivity are needed.

In general, by arranging several elementary radiating elements into an array a directive beam of radiation can be obtained, resulting in an antenna with higher gain. This permits the total transmitted power to be reduced considerably for the same signal strength at the receiving site and, of course, to decrease "radiation pollution."

In order to present the general method used to analyze arrays, let us consider the general array shown in Figure 7.4. This array consists of N elements. The radiation vector of this structure is given as the relation

$$\overline{N} = \Sigma \overline{N}_m \cdot \exp(jkr_m \cdot \cos \chi_m) \qquad (7.28)$$

where $\overline{N}_m$ denotes the radiation vector of the mth element, k is the wave number

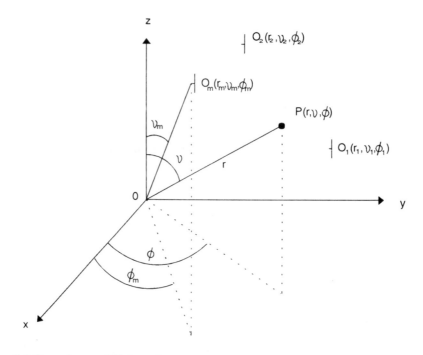

Figure 7.4 General array of N elements.

$$\cos \chi_m = \cos \upsilon_m \cdot \cos \upsilon + \sin \upsilon_m \cdot \sin \upsilon \cdot \cos(\phi_m - \phi) \qquad (7.29)$$

where r_m, υ_m, and ϕ_m are the spherical coordinates of the position of the mth element while r, υ, and ϕ are the coordinates of the position where the radiation field is expressed.

In the case where the elementary elements are identical, have the same orientation, and are excited with relative amplitude C_m and phase α_m for the mth element, the electric field of the array is expressed as follows:

$$\overline{E}(r, \upsilon, \phi) = \bar{f}(\upsilon, \phi) \cdot \frac{\exp(-jkr)}{4\pi r} \sum_{m=1}^{N} C_m \exp(j\alpha_m + jkr_m \cos \chi_m) \qquad (7.30)$$

where $\bar{f}(\upsilon, \phi)$ is the electric-field radiation pattern of the elementary antenna used in the array.

With respect to the array's input impedance, the following system of equations is used:

$$V_1 = Z_{11}I_1 + Z_{12}I_2 + \cdots + Z_{1N}I_N$$
$$V_2 = Z_{21}I_1 + Z_{22}I_2 + \cdots + Z_{2N}I_N \qquad (7.31)$$
$$\cdots$$
$$V_N = Z_{N1}I_1 + Z_{N2}I_2 + \cdots + Z_{NN}I_N$$

where V_m denotes the voltage in the center of the mth element, I_m its input current, Z_{mm} is the self-impedance of the mth element and finally, Z_{mn} is the mutual impedance between the mth and nth elements. These relations, together with the well-known transmission line equations, are used in order to obtain the input impedance of any array. For a detailed description of the array of elements, the reader is referred to many textbooks [1,2,3,4,18].

The relatively small size of VHF (450 MHz), UHF (900 MHz), and especially 1.7 GHz arrays, results in a wide range of construction solutions for in-house use. Yagi dimensions of several configurations for 432 MHz are described in many references [2].

Apart from the Yagi and Quagi configurations [1,2], the corner reflector antenna may prove to be a very useful antenna for in-building coverage (e.g., it can be placed in one of the upper corners of a large room in order to avoid interference from outside) (see Fig. 7.5). The feeder antenna is usually a $\lambda/2$ dipole (Section 7.2).

Generally when a single driven element is used the reflector screen may be bent to form an angle, giving an improvement in the radiation pattern and gain. At 450 MHz the reflector size assumes practical proportions and at 900 MHz or higher frequencies practical reflectors can approach ideal dimensions (i.e., very large in terms of number of wavelengths, resulting in more gain and probability in sharper patterns).

There are advantages of this simple type of antenna system. It is simple to construct; it is broadband; and it has gain from 10–15 dB, which is quite convenient for in-house communications. Also, the orientation of this type of antenna system is quite easy. The corner angle can be 90°, 60°, or 45°, but the minimum side length must be 0.25λ, 0.35λ and 0.5λ, respectively. The radiation resistance of the corner reflector is a function of the spacing between the feed and the apex of the reflector and can take values from 20 to 100 for spacings from 0.2λ to 0.4λ, approximately (for corner angle of 90°) [2].

The corner reflector antennas can be constructed efficiently for a wide class of frequencies, starting from VHF up to 25 GHz.

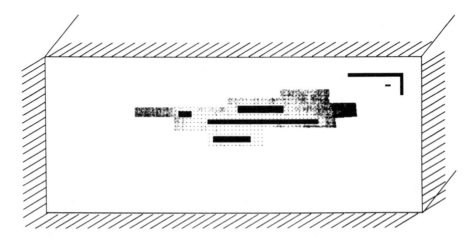

Figure 7.5 Coverage of a corner reflector antenna.

Depending on the degree of mobility, directional antennas at the mobile unit can be effective, especially for applications where the user is not moving when using the equipment. As an example, a microstrip-phased array with considerable diversity may be used.

The orientation of such a configuration can be controlled electronically. A possible configuration is shown in Figure 7.6 where the functional entity "controller" controls the phases of the currents of the array elements (e.g., Motorola's 18-GHz Altair).

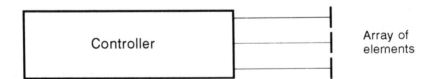

Figure 7.6 A configuration of directional antennas electronically controlled.

Notice that such an antenna system is quite expensive. However there are situations where it can be used, especially under the following circumstances: noisy environments, very high bit rates, high reliability, and low mobility of the user's equipment.

7.4 DIVERSITY RECEPTION

7.4.1 General

Conventional antennas have been used in the past for most indoor communication systems. One of the most serious problems for the indoor systems is that the communication is interrupted when the receiving antenna is located at the minimal areas of standing wave distribution of the electric field.

In general, two major problems have impact on the quality of the signal reception in a wireless environment. One is related to long-term fading caused by shadowing, and the other is related to the short-term fading caused by multipath propagation. In an indoor environment the first one does not seem to have high priority, while the second one is the most critical because of the nature of the inhouse environment (multipath propagation).

The worst case would be the blockage of the direct wave between the network antenna and the mobile unit antenna by objects that interfere temporarily or permanently. In such cases the scattered waves arrive from 360° around the terminal, resulting in a highly dependent signal strength. The microscopic diversity applied in the terminal would be a potential solution for such a problem without increasing the radiating power of the system.

In Table 7.1 the different diversity schemes are presented. When using diversity reception, the main changes may be to the required bandwidth and to the antenna size. The influence of the selected diversity scheme to these characteristics is approximately shown in Table 7.1.

For completeness, various microscopic diversity schemes are discussed briefly in the following sections.

7.4.1.1 Space Diversity

In the case of small spacing between the elements of the antenna, the received signal fading by these elements are less correlated. The 0.25λ spacing is a good

Table 7.1
Diversity Schemes

Diversity Type	Bandwidth Extension	Antenna Size
Space	Zero	Increases
Frequency	Double	No change
Polarization	Zero	Increases
Field	Zero	Increases slightly
Time	Double	No change

rule, although it is not the only one. At frequencies above 1 GHz, this spacing is less than 7.5 cm and seems to be practical. A better rule can be derived by measurements inside buildings, but it depends on the operating frequency and the type of elements (dipoles, microstrip, etc.).

7.4.1.2 Frequency Diversity

It is suggested that frequency diversity not be used for wireless high-speed LANs because of the expected limitations in the system bandwidth. In contrast, it may be used in low-speed wireless LANs and wireless phones.

7.4.1.3 Polarization Diversity

The use of two polarization components (usually with vertical polarizations) can result from the provision of two uncorrelated signal fadings at the receiving site. One possible drawback of using this method is the 3-dB power reduction at the transmitting site due to splitting the power into two different polarized antennas.

7.4.1.4 Field Component Diversity

This concept seems to be promising for the in-house environment. It is based on the fact that when an E field is propagating, the H field is always associated with it and both fields carry the same information message. This scheme does not require physical separation of the two antennas as in space diversity, so it needs less available space. A possible configuration based on field diversity is described in next paragraph. It is suggested that configurations based on this concept be studied in the future, both theoretically and experimentally.

7.4.1.5 Time Diversity

It is suggested that this not to be taken into account because in an in-house wireless environment it does not solve the multipath fading problem, especially when the receiver is at the same place for some time. This situation is very usual in every scenario. On the other hand, it needs more bandwidth available for a system based on time diversity reception.

Generally speaking, the diversity reception tends to reduce the depth of fades on the combined output, provides improved equipment reliability, and finally, the combined output signal-to-noise ratio is improved over that of any single path. In Figure 7.7 a simplified receiving system based on diversity reception is presented. Note that the N receiving antennas can be either separated in space, or may have

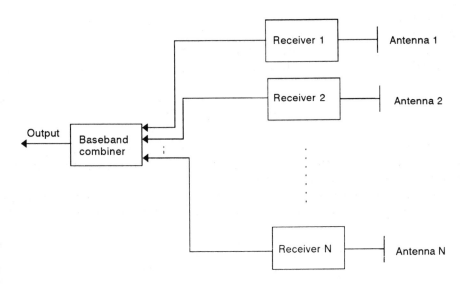

Figure 7.7 Simplified block diagram of diversity reception.

different polarization, or they are of different nature (e.g., one is a slot and the second one is a microstrip). The signals received by each antenna are translated in the baseband frequency (or at least at the intermediate frequency) and then they are combined in a separate stage, giving the combined output.

7.4.2 Combining Techniques

Generally, the combining techniques discussed in the following sections may be used.

7.4.2.1 Selective Combining

When this combining scheme is in use, the selection combiner uses the receiver with the maximum output at the time. The performance analysis of each combiner is carried out in terms of signal-to-noise ratio. In an in-house environment the probability density function of the SNR can be written as follows:

$$p(s) = 1/s_0 \exp(-s/s_0) \qquad (7.32)$$

where s_0 is the ratio of the mean carrier power divided by the mean noise power.

Note that this model is valid under the assumption that the signal is Rayleigh distributed and is received in the presence of Gaussian noise.

Then, the cumulative probability distribution of the signal with maximum strength that is taken from the N diversity branches (Fig. 7.7) can be easily expressed as follows:

$$p_N(s) = [1 - \exp(-s/s_0)]^N \qquad (7.33)$$

In order to obtain (7.33), the local mean-squared values of the signals are assumed to be independent while the noise in each diversity branch is assumed to be additive and independent of the signal.

As the cumulative probability distribution is known, it is a simple matter to obtain the mean signal to noise ratio at the output of the combiner as follows:

$$(S/N)_{\text{mean}} = \int_0^\infty s \, \frac{dP_N(s)}{ds} \, ds = s_0 \sum_{i=1}^{N} \frac{1}{i} \qquad (7.34)$$

From the above-obtained relation an improvement to the mean signal-to-noise ratio by a factor 1.5 in the case of two diversity branches is observed, at least theoretically.

A slightly different system that is more practical than the one described above is the so-called scanning diversity system. When this system is used there is no attempt to find the best branch, but one that is acceptable (i.e., above a predetermined threshold).

7.4.2.2 Maximal Ratio Combining

When this combining technique is used, each branch signal is weighted by a coefficient w_i. On one hand, the signal in the output of the combiner has the following form:

$$r_{\text{total}} = \sum_{i=1}^{N} w_i \cdot r_i \qquad (7.35)$$

where r_i is the signal at the ith branch.

On the other hand, the total noise at the output of the combiner is given as the relation:

$$n_{\text{total}} = n \cdot \sum_{i=1}^{N} w_i^2 \qquad (7.36)$$

where n represents the noise of the indoor communication channel. Using the above relations we can obtain the resulting signal-to-noise ratio at the output of the combiner as follows:

$$(S/N)_{\text{total}} = \frac{(r_{\text{total}})^2}{2 \cdot n_{\text{total}}} \tag{7.37}$$

It has been proven that the signal-to-noise ratio is maximized when the weighting factors w_i are selected to be equal to the ratio r_i/n, and then the signal-to-noise ratio at the output of the combiner takes the following form:

$$(S/N)_{\text{total}} = \sum_{i=1}^{N} (S/N)_i \tag{7.38}$$

where $(S/N)_i$ is the signal-to-noise ratio in each branch.

Then, the mean *SNR* at the output of the combiner is obtained as follows:

$$(S/N)_{\text{mean}} = N \cdot s_0 \tag{7.39}$$

where again s_0 represents the ratio of the mean carrier power divided by the mean noise power. From the above-obtained relation an improvement to the mean signal-to-noise ratio by a factor of 2 in the case of two diversity branches is observed, at least theoretically.

7.4.2.3 Equal Gain Combining

The equal gain combiner simply adds the signals of diversity branches, so its principle is similar to maximal ratio combining but the weighting factors are equal to one. In this case the mean signal-to-noise ratio at the output of the combiner is related to the mean carrier power divided by the mean noise power (s_0) by the relation

$$(S/N)_{\text{mean}} = s_0[1 + (N - 1)\pi/4] \tag{7.40}$$

Again, from the above relation an improvement is observed to the mean signal-to-noise ratio by approximately a factor of 1.785 in the case of two diversity branches.

7.4.3 Examples of Diversity Antennas

7.4.3.1 Diversity Antenna System Composed From Two Monopoles Above Ground

A diversity antenna system consists of a ground plane and two λ/4 monopoles above it, as shown in Figure 7.8. The two signals from the two monopoles are combined using a T balanced junction so that equal gain combining is employed.

In order to measure the influence of the space diversity of antenna systems in the in-house environment, measurements inside a typical corridor of a large building have been carried out at 900 MHz. A sample of the received signal levels are shown in Figures 7.9 and 7.10. Figure 7.9 corresponds to one monopole-receiving antenna, while Figure 7.10 corresponds to a space diversity scheme with two monopoles and spacing λ/2. Five hundred signal samples every 1 second have been taken and during this time the receiving antenna system was moving with a mean velocity 0.02 m/sec. Comparing these two figures, the advantage of using space diversity is observed.

Also, 50 static measurements were carried out every 0.5 m for a total distance of 85m. For each sample the mean value and the standard deviation were computed

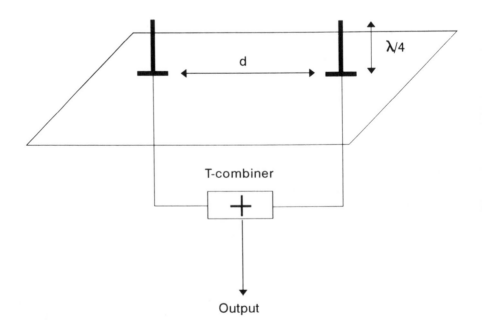

Figure 7.8 Configuration of a simple space diversity antenna system.

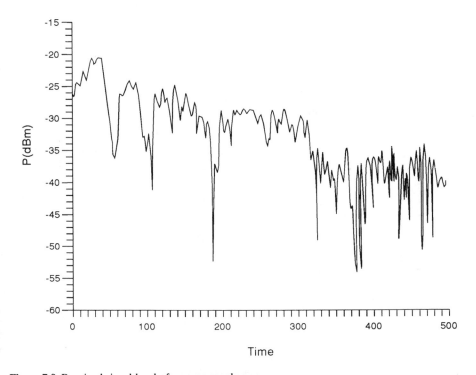

Figure 7.9 Received signal level of one monopole.

and the results are shown in Figures 7.11 and 7.12 for the simple monopole and in Figures 7.13 and 7.14 for the space diversity system. From these figures it can be observed that the space diversity system reduces the effect of multipath fading.

7.4.3.2 A Portable Antenna Consisting of a Whip and Microstrip

In 1986, a study of the possibilities of using antenna diversity for portable telephones (900-MHz band) was carried out at the Technical Research Center of Finland [11]. A whip antenna was used for both transmitting and receiving while a microstrip antenna operating at a quarter wavelength resonance above ground plane was the second branch for diversity reception. This configuration works as a combination of space, pattern, and polarization diversity.

The configuration was found to work quite well, therefore, for the cases that are similar to cellular telephony. This antenna system can be used in wireless LANs but, of course, its actual performance in the indoor environment must be inves-

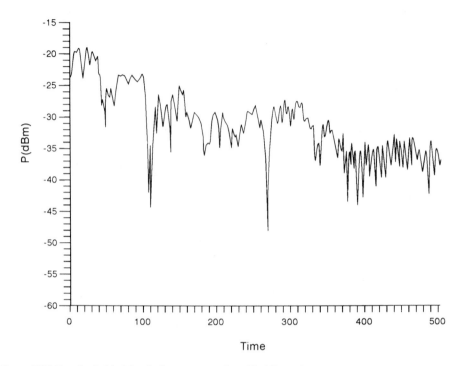

Figure 7.10 Received signal level of two monopoles with λ/2 spacing.

tigated. Also, based on the same principle, similar antenna systems for other frequencies may prove to be quite efficient for use in wireless LANs.

7.4.3.3 A Small Printed Antenna Composed of a Slot and a Wide Strip

One of the biggest problems for indoor transmission systems is that the communication is sometimes interrupted when a receiving antenna happens to be located at the minimal areas of a standing wave distribution of electric fields.

In an indoor communication environment, the electromagnetic field distribution may suffer from multiple reflections and scattering of the transmitted waves, resulting in standing wave distribution or polarization rotation. The antenna presented in [7] overcomes the standing waves, as it is based on the reception of the transmitted signals through both the magnetic and electric fields.

One possible configuration based on the above principles is an antenna system composed of a half wavelength slot and a wide-strip dipole. The radiation patterns of these antennae are given in (7.9), (7.16) and (7.17), respectively. A passive

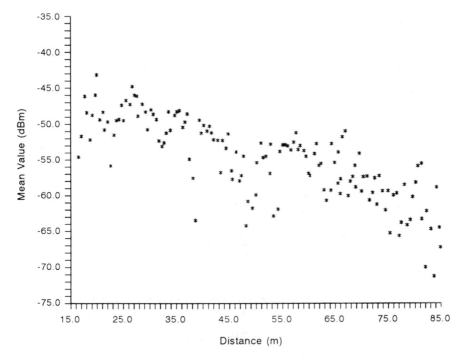

Figure 7.11 Mean value *m* of the received signal by one monopole.

signal combiner may be used as a simple combining mechanism in this diversity scheme. The slot and the wide strip may be made on both sides of a printed circuit board and the element spacing can be made as small as possible, resulting in a rather small-size antenna system. The antenna presented in [7] has been developed for an operation frequency of 320 MHz and its validity was checked by experiments.

For operation in other frequencies, a good rule, at least theoretically, is to use the frequency translation transformations [2]. It is suggested that an antenna based on the same principle for different frequencies may prove to be a rather efficient reception system for wireless LANs applications.

7.4.3.4 A Portable Data Terminal

Although the antenna system that is presented in this section is not a direct application of diversity, it is a good example of an antenna system that is based on ideas presented in this chapter. The portable data terminals have radiation requirements somewhat different from those of traditional portable phones. In the radiation

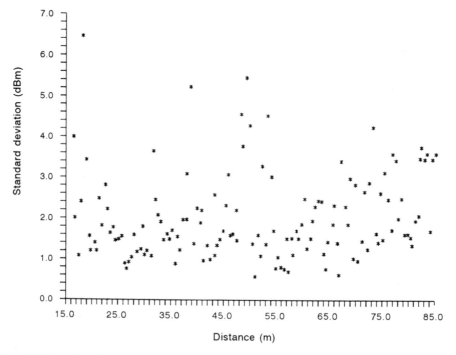

Figure 7.12 Standard deviation of the received signal by one monopole.

environment the objects in the immediate vicinity of the radiator can be, in one sense, predictable for mobile phones. For example, the antenna of a mobile phone could be in front of the face of a person with a few meters of free space around the head of the user. For a portable data terminal the situation is quite different.

In general, the data terminal is expected to radiate from any position in the close vicinity of conducting or nonconducting surfaces (e.g., on a metal slab of a table, inside a briefcase, etc.). An antenna system meeting such requirements is described in [10].

The antenna developed for this specific application is a transmission-line radiator $\lambda/2$ long (operating at a frequency of 800 MHz). The antenna has four vertical radiators. Only one is actively fed by the transmitter (Fig. 7.15). The other three are excited by coupling through the transmission line.

The two vertical elements at each end are so tightly coupled that each pair radiates as a single short dipole (vertical field polarization). The transmission line is unbalanced and radiates an electric field with orthogonal polarization with the one radiated by the vertical radiators. This polarization is radiated even if the radio

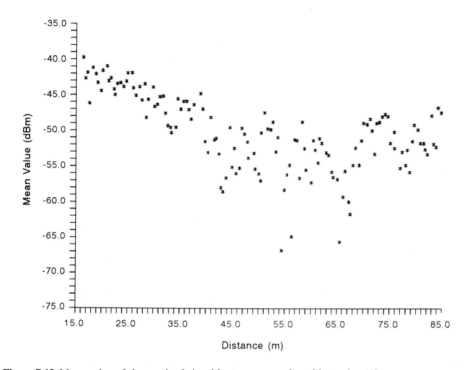

Figure 7.13 Mean value of the received signal by two monopoles with spacing λ/2.

case is laid flat on a conducting surface, such as a metal table. In this respect this antenna system is based partly on polarization diversity.

In conclusion, the antenna presented in [10] is totally enclosed within the outer plastic of a portable data terminal. Because of its particular construction the antenna is capable of efficient radiation in both freespace conditions and when the radio is laid flat on a conducting surface.

7.4.3.5 An Intelligent Six-Sector Microwave Antenna

Motorola's wireless in-building network uses low-power microwave transmission at 18 GHz. An intelligent six-sector microwave antenna [21] has been designed to overcome the multipath distortion problem. Each transmission of data takes the best of numerous paths made possible by this antenna. Multiple path options maintain maximum signal gain and minimize multipath distortion. During data reception the antenna of the receiving site samples the incoming signal from the

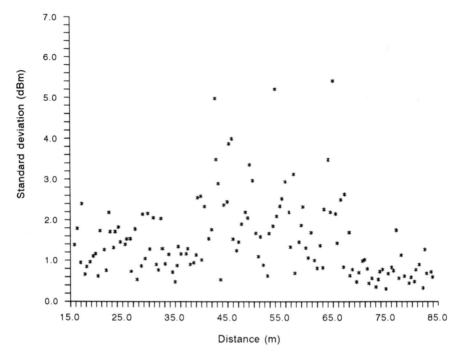

Figure 7.14 Standard deviation of the received signal by two monopoles with spacing λ/2.

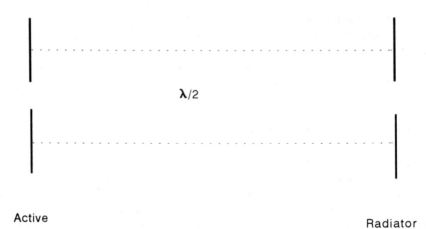

Figure 7.15 Configuration of a portable data terminal.

six transmitting antennae of the transmitting side for each of its six receiving sectors. The intelligent antenna automatically and continuously selects the best incoming signals. Identities of the transmitting sectors are encoded in the signals themselves so the receiving antenna can remove duplicate signals from other sectors. To reach this goal Motorola uses a single-chip radio frequency digital-signal processor for data synthesis.

7.5 INFLUENCE OF THE INDOOR ENVIRONMENT TO THE ANTENNA PERFORMANCE

Up to this point a theoretical presentation of all the possible antenna systems for both the network and the terminal equipment for in-house environments has been given. In this survey, some comments and hints for the efficient use of each antenna system for indoor environments were also given. In this section the performance of the transmitting/receiving antenna system is discussed in the real in-house environment, because it is expected that the wireless LANs will be used especially in office or business environments.

7.5.1 Safety Aspects

First, the radiation safety aspect has to be taken into account. This in general means that the radiation exposure levels must be kept to a minimum, as large amounts of RF energy may cause serious heating problems.

In 1982, the American National Standards Institute (ANSI) released a standard for the limits of RF radiation [19]. With respect to wireless LANs, two aspects have to be taken into account. The RF exposure limit is 5 mW/cm^2 above 1,500 MHz, while it rises from 1 mW/cm^2 at 300 MHz to 5 mW/cm^2 at 1,500 MHz. As the operation frequency in wireless LANs is expected to be above 1,700 MHz (mobile telephony of next generation systems), the exposure limit must be taken to be 5 mW/cm^2. Second, the user of a wireless LAN system will use the equipment not temporarily, but for long periods of time (although actual transmission may be in short bursts). For this reason, the RF exposure limits must be kept to the minimum level. During the installation of the system, measurements must be made to guarantee the safety with respect to radiation level. An approximate calculation of the power density in the far field region of an antenna may be based on the relationship

$$P = (W \cdot G)/(4 \cdot \pi \cdot r^2) \qquad (7.41)$$

where W is the average power at the transmitting antenna feed, G is the antenna gain, and r represents the distance between the antenna and the observation point.

7.5.2 Statistical Model

The influence of the human body and generally of the in-house environment on the performance of the terminal antenna system reflects the dependence of the S_{11} coefficient on the relative position of the terminal antenna. The reason for developing a model for the prediction of the probability that S_{11} is less than a predefined value is that it can be used as a comparison tool between different antenna types for the terminal in the in-house environment.

The statistics of the antenna reflection coefficient may be studied using the model [20]

$$P(S_{11} < p) = 1 - a \cdot \exp(-ub)/p^b \tag{7.42}$$

for $p > \epsilon$, where $P(S_{11} < p)$ represents the cumulative intensity distribution of the reflection coefficient and a, b, u, and ϵ are parameters depending on the specific antenna type and its construction characteristics. The parameters a, b, u, and ϵ, which characterize the whole cumulative intensity distribution are not independent of each other, but they should satisfy the following relationship

$$a = \exp(u \cdot \epsilon) \cdot \epsilon^b \tag{7.43}$$

as a result of the obvious relation $P(S_{11} < \epsilon) = 0$.

The probability density function associated with (7.1) is

$$f(p) = \begin{cases} a \cdot \exp(-up)(u + b/p)/p^b & p > \epsilon \\ 0 & p < \epsilon \end{cases} \tag{7.44}$$

As a direct consequence of the definitions of the initial moments, after a straightforward algebra, we can obtain

$$E(p) = \epsilon + a \cdot u^{b-1} \Gamma(1 - b, u\epsilon) \tag{7.45}$$

$$\begin{aligned} \text{Var}(p) = {} & a^2 u^{2(b-1)} \Gamma^2(1 - b, u\epsilon) \\ & + 2au^{b-2}[\Gamma(2 - b, u\epsilon) - \epsilon u \Gamma(1 - b, u\epsilon)] \end{aligned} \tag{7.46}$$

where $E(p)$ is the mean value, $\text{Var}(p)$ is the variance and $\Gamma(a, x)$ represents an incomplete Gamma function. These two relations can be used in order to determine the theoretical values of the mean and the variance of the antenna's reflection coefficient.

In order to obtain the parameters a, b, ϵ, and u, a set of measurements when moving the terminal antenna near the user are taken, and then the parameters are calculated using the least-square method. In Figures 7.16 and 7.17 both measurements and theoretical values of $P(S_{11} < p)$ are shown for one monopole $\lambda/4$ above ground and two monopoles with spacing $\lambda/2$ respectively. In those figures the continuous line represents the measurements while the asterisks (*) represent the theoretical curves. Note that the simple diversity scheme composed of the two monopoles has better performance with respect to its reflection coefficient in the real environment than the simple monopole antenna.

The following comparison method with respect to the variation of the antenna's reflection coefficient may be based on the above-developed model. This method includes the following steps.

1. Take measurements of the antenna's S_{11} parameter, moving the antenna near the user's body. Notice that the same sample must be taken for all antennas under study.
2. Calculate the parameters a, b, u, and ϵ of the model.
3. Calculate $E(p)$ and $\text{Var}(p)$ from (7.45) and (7.46).
4. The antenna with the smallest $\text{Var}(p)/E(p)$ ratio will have the best performance with respect to the variation of the S_{11} parameter in the in-house environment.

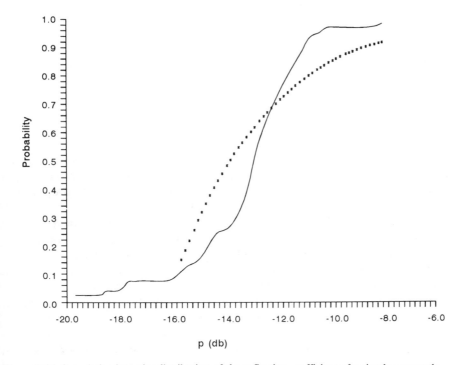

Figure 7.16 Cumulative intensity distribution of the reflection coefficient of a simple monopole.

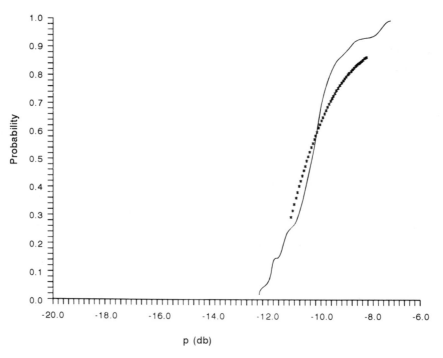

Figure 7.17 Cumulative intensity distribution of the reflection coefficient of two monopoles with spacing λ/2.

7.6 CONCLUSIONS

In this chapter, existing possible antenna systems for both the network and the terminal for wireless LANs applications have been examined. Special attention has been given to the terminal antenna in order to overcome the multipath propagation in the in-house environment. Simple antennas have been described because they can be used either individually or in combination (arrays, diversity antenna systems) to form an efficient transmitting/receiving system for wireless LANs applications. Also, the principles of directional antennas have been presented because, depending on the coverage area, in many cases in-house wireless communications antenna systems with considerable directivity may be used. Depending on the degree of mobility, directional antennas can be also be used effectively at the mobile terminal, especially for applications where the user is not moving when the equipment is used. Printed antennas have also been described as they can be used in both the network and the terminal because of the serious advantages they offer. Special attention has been paid to the examination of using diversity techniques at the receiver side in order to overcome the indoor multipath fading. Some examples of antenna systems based on diversity techniques have been presented, while some measurements at 900 MHz have been presented in order to show the advantages of even simple space-diversity schemes in comparison to simple antennas. In the final part of this chapter, the influence of the human body on the antenna's performance has been examined through the variation of the antenna's reflection coefficient when the antenna system is inside the real office environment. In the future, the development of antenna systems for indoor applications might be directed towards using more sophisticated antennas based on the ideas presented in Section 7.5.

REFERENCES

[1] Ramo, S., J. R. Whinnery, and T. Van Duzer. *Fields and Waves in Communication Electronics*, New York: John Wiley & Sons, Inc., 1965.

[2] W. L. Weeks. *Antenna Engineering*, New York: McGraw-Hill, 1968, pp. 62–84.

[3] Rufdge, A. W., K. Milne, A. D. Olver, and P. Knight. *Handbook of Antenna Design, Vol. 1*, IEEE, 1982.

[4] American Radio Relay League, *The APRL Antenna Handbook*, Newington, CT: 1988, §§1.16–1.20, 8.1–8.22.

[5] Parsons, J. D., and J. G. Gardiner. *Mobile Communication Systems*, Glasgow: Blackie and Son, pp. 189–240.

[6] Lee, W.C.Y. *Mobile Communications Design Fundamentals*, Indianapolis: Howard W. Sams & Co., 1986, pp. 49–63.

[7] Rappaport, T. S. "Indoor Radio Communications for Factories of the Future," *IEEE Communications Magazine*, Vol. 27, No 5, May 1989, pp. 15–24.

[8] Cox, D. C. "Portable Digital Radio Communications—An Approach to Etherless Access," *IEEE Communications Magazine*, Vol. 27, No 7, July 1989.

[9] Ito, K., and S. Sasaki. "A Small Printed Antenna Composed of Slot and Wide Strip for Indoor Communication Systems," *IEEE Antennas and Propagation Symposium*, June 12, 1988, pp. 716–719.

[10] Kishk, A. A., H. A. Auda, and B. C. Ahn. "Radiation Characteristics of Cylindrical Dielectric Resonator Antennas with New Applications," *IEEE Antennas and Propagation Society Newsletter*, Feb. 1989, pp. 7–16.

[11] Bahl, I. J. *Microstrip Antennas*, Norwood, MA: Artech House Inc., 1980, pp. 2–9, 32–56.

[12] Garay, O., and Q. Balzano. "An Antenna for a Portable Data Terminal," *IEEE Conference*, 1984, pp. 54–56.

[13] EEC COST Project 207, *Digital Land Mobile Radio Communications*, 1988.

[14] Bhartia, P., K.V.S. Rao, and R. S. Tomar. *Millimiter-Wave Microstrip and Printed Circuit Antennas*, Norwood, MA: Artech House Inc., 1990, pp. 9–19.

[15] Nakaoka, K., K. Itoh, and T. Matsumoto. *Microstrip Line Array Antenna and its Application*, International Symposium on Antennas and Propagation (Japan), 1978, pp. 61–64.

[16] Compton, R. T., and R. E. Collin. *Slot Antennas*, in *Antenna Theory*, Pt. 1, New York: McGraw-Hill Book Co. 1969.

[17] James, J. R., P. S. Hall, and C. Wood. *Microstrip Antenna Theory and Design*, London: IEE Peter Peregninus, 1981.

[18] R.E. Collin, *Antennas and Radio wave Propagation*, McGraw-Hill Book Co., 1985, pp. 107–116.

[19] American National Standards Institute, "Safety Levels with Respect to Human Exposure to Radio Frequency Electromagnetic Fields," ANSI C95.1-1982.

[20] Capsalis, C. N., and J. D. Kanellopoulos. "Prediction Method for the Rain Attenuation Statistics Based on a Global Rainfall-rate Distribution Model," *Anntenna Telecommunications*, Vol. 43, No. 9–10, 1988, pp. 528–533.

[21] Freeburg, T. A. "Enabling Technologies for Wireless In Building Network Communications—Four Technical Challenges, Four Solutions," *IEEE Communications Magazine*, April 1991, pp. 56–64.

Chapter 8

Coding And Modulation Techniques For RF LANs

Cyril J. Burkley
University of Limerick
Plassey Technological Park, Limerick, Ireland

8.1 INTRODUCTION

Mobile communications is the fastest growing sector of the communications industry, with cellular and cordless telephones and pagers approaching everyday usage by a large number of people. The extension of mobile communications to the business and industrial environments is obviously the next stage of development. Wireless networks will provide user mobility and flexibility and will reduce installation costs.

A wireless communications network system is composed of a number of mobile transceiver units and a fixed base station. For a wireless local area network, the varying propagation and fading conditions that are present in the radio channel impose constraints on the choice of technology that can be used. This chapter concentrates on the implementation technology for such systems and reviews the existing state-of-the-art technologies for wireless communications for RF frequencies. Various aspects of transmitter and receiver technologies are reviewed, including modulation and coding, access schemes, detection, synchronization, and equalization techniques.

The current telecommunications network is becoming more fully digital and therefore future wireless communications systems will inevitably also use digital techniques. Digital systems offer several important advantages, including significantly improved performance, lower costs, possibility of error detection and correction, more flexible services, and the integration of voice and data. Consequently, digital systems are becoming more and more popular and future wireless communication systems inevitably also will use digital techniques.

8.2 DIGITAL MODULATION TECHNIQUES

As with any digital or data transmission system, there are several criteria that must be considered when choosing a modulation technique for a digital wireless/cellular system. The most important criteria are:

1. Spectral efficiency, in order to maximize the number of channels per MHz of bandwidth,
2. Bit error performance, both in noise and against cochannel interference,
3. Applicability to the cellular environment, and
4. The ease and cost of implementation.

Because of the limited spectrum available for wireless systems, the spectral efficiency of any proposed new system is probably the most important consideration.

There are many different modulation techniques that can be considered for digital wireless systems. For an in-house network, the varying fading conditions that are present in the wireless channel impose additional constraints on the modulation choice. Digital modulation techniques can be divided into two broad classes: constant-envelope modulation techniques and linear modulation techniques. Constant-envelope modulation systems allow transmitter amplifiers to operate in a nonlinear class C mode, resulting in higher efficiencies and lower implementation costs. However, these are achieved at the expense of spectral efficiency and constant envelope schemes are limited to a spectral efficiency of approximately 1 bit/s/Hz, regardless of the number of modulation levels used. Linear modulation techniques, which will have a varying transmitted envelope, require a linear transmitter amplifier with a resulting increase in cost and complexity. However, the modulations yield much better spectral efficiency, which may justify the additional costs.

8.2.1 Phase Shift Keying

The simplest and most commonly used digital modulation technique is phase shift keying (PSK). A PSK signal is generated by applying the carrier waveform $A \cdot \cos w_c t$ to a balanced modulator and applying the binary data signal $b(t)$ as the modulating signal, as shown in Figure 8.1. If the binary data is coded in a form such that a "1" or mark waveform is represented by $b(t) = +1$ and an "0" or space waveform is represented by $b(t) = -1$, then the PSK signal is represented by

$$v_{PSK}(t) = \pm A \cos w_c \cdot t \tag{8.1}$$

Thus the output waveform is a constant amplitude signal, which alternates between two different phase states, 0 deg and 180 deg.

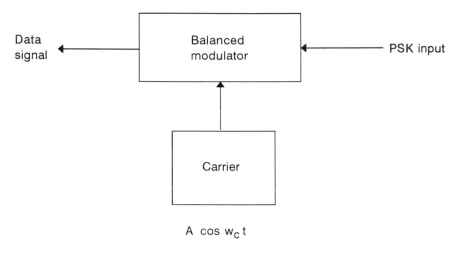

Figure 8.1 The PSK modulator.

Demodulation at the receiver can be achieved by performing the same process as at the transmitter, but coherent demodulation is necessary for satisfactory performance in the additive white Gaussian noise (AWGN) channel. The block diagram of the coherent PSK demodulator is shown in Figure 8.2.

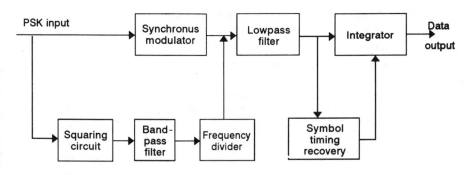

Figure 8.2 The PSK demodulator.

For coherent demodulation the carrier at the correct phase must be extracted at the receiver and this can be accomplished by squaring the received signal, filtering out the desired second harmonic with a bandpass filter, and dividing the resultant frequency by two. If $v(t) = \pm A \cdot \cos(W_c t + \Theta)$, where Θ is the phase shift corresponding to the delay between the transmitter and the receiver, then the squared signal can be expressed as

$$v^2(t) = A^2 \cdot \cos^2(w_c t + \Theta) = \frac{1}{2} \cdot [A^2 + A^2 \cdot \cos 2(w_c t + \Theta)] \qquad (8.2)$$

and the divide-by-two circuit yields $\cos(W_c t + \Theta)$, which is the desired carrier. The recovered carrier is then multiplied with the received signal to generate

$$\pm A \cdot \cos(w_c t + \Theta) \cdot \cos(w_c t + \Theta) = \pm \frac{A}{2} \cdot [1 + \cos 2(w_c t + \Theta)] \qquad (8.3)$$

and this signal is then applied to an integrator. The desired output signal is the integrator output at the end of the bit interval, which is determined by the symbol timing recovery circuit.

8.2.2 Quadrature Phase Shift Keying

To achieve higher bit rates, four-phase modulation is used in which each symbol or interval carries two data bits. Such a system, which is referred to as quadrature PSK (QPSK), provides twice the data throughput in the same bandwidth as compared to PSK. At the transmitter, a serial-to-parallel converter is used to provide pairs of binary digits that modulate two carriers that are offset from each other by 90 deg. The two output signals from the modulators are summed to provide a transmitted signal, which alternates between four different phase states, -90 deg, 0 deg, 90 deg, and 180 deg. A block diagram of the quadrature modulator is shown in Figure 8.3.

A receiver for the QPSK signal is shown in Figure 8.4. Synchronous demodulation is required and hence it is necessary to regenerate the carrier signals $\cos W_c t$ and $\cos(W_c t + 90)$ at the receiver. The received QPSK signal is also applied to the two synchronous demodulators, each consisting of a balanced modulator and an integrator, as before. However, in this case the integrators integrate over a two-bit period and a parallel-to-serial converter performs the process of converting the two-bit symbol back to two bits of serial data. As before, a bit synchronizer is required to establish the beginnings and ends of the bit intervals so that times of integration can be established.

The amplitude spectrum of PSK is wide and decays slowly with offset from the center frequency [1]. Pulse shaping can be used to reduce the bandwidth of the transmitted signal while keeping the intersymbol interference to a minimum.

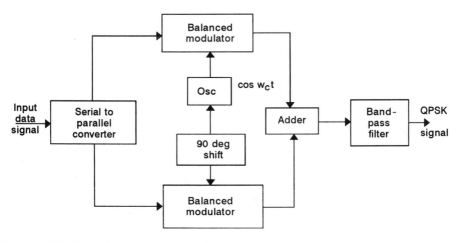

Figure 8.3 QPSK modulator.

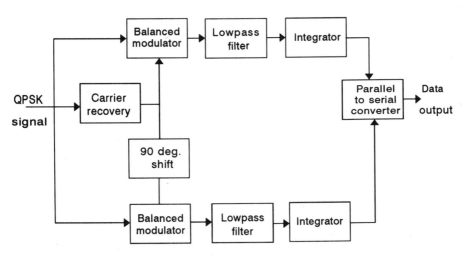

Figure 8.4 QPSK demodulator.

However, once pulse shaping is used it becomes a linear modulation technique and linear amplification must be used to preserve the pulse shape. If nonlinear power amplifiers are used, the pulse shape will be destroyed and the signal bandwidth

---

will spread. Nyquist pulse shaping using raised cosine pulses is employed to provide high spectral efficiency [2].

8.2.3 Multilevel PSK Systems

In PSK each bit is transmitted individually, while in QPSK two bits are combined and transmitted as one of four possible phases. The scheme can be extended. If N bits are combined, then in this N-bit symbol there are $M(= 2^N)$ possible symbols. Such systems are referred to as M-ary PSK systems. In PSK, QPSK, and M-ary PSK the transmitted signals are distinguished from one another in phase but they are all of the same amplitude. In quadrature amplitude modulation (QAM) the signals are allowed to differ not only in phase but also in amplitude [3].

Linear modulation schemes such as shaped PSK and QAM, which use linear power amplifiers, can achieve efficiencies greater than 1 bit/s/Hz, with the efficiency increasing with the number of levels used. For example, the use of a four-level modulation scheme allows the transmission of two bits per symbol with no degradation (on a S/N per bit basis) over binary modulation because the bits are sent in quadrature on independent carriers. Doubling the number of bits per symbol doubles the data rate that is possible in a given bandwidth. However, the increase in the number of modulation levels leads to an increase in the bit error rate due to the reduced distance between the elements of the signal set. The use of linear amplifiers, which provide better out-of-band radiation performance, further increases the spectral efficiency of the system.

Because of the need for linear RF amplifiers, which are still quite costly, linear modulation techniques have not received as much attention as constant envelope schemes. The most important linear modulation methods are based on various forms of phase shift keying, particularly differential PSK, QPSK, offset QPSK, and M-ary PSK [4,5].

8.2.4 Frequency Shift Keying

Frequency shift keying (FSK) is a constant-envelope modulation scheme in which the carrier frequency is shifted from a mark frequency (corresponding to a binary 1) to a space frequency (corresponding to a binary 0) according to the baseband data signal. It is identical to modulating an FM carrier with a binary digital signal. Thus, the transmitted signal is either

$$v(t) = \begin{cases} A \cdot \cos w_1 t & \text{for ``1''} \\ A \cdot \cos w_0 t & \text{for ``0''} \end{cases} \tag{8.4}$$

Frequency modulation schemes are characterized by their modulation index, which is defined as the ratio of the peak frequency deviation to the highest frequency component in the modulating signal. An FSK signal can be generated either by switching the transmitter output line between two different oscillators or by feeding the data signal into a frequency modulator, as shown in Figure 8.5.

FSK signals can be demodulated by using either noncoherent detection or synchronous detection, as shown in Figure 8.6. With noncoherent detection, the received signal is applied to two bandpass filters that have a narrow passband so that complete rejection of undesirable frequencies can be achieved. The filter outputs are applied to envelope detectors, whose outputs are compared by a comparator. The comparator generates a binary output, whose level depends on which input is larger. The synchronous or coherent detector uses two product detectors to determine which transmitted frequency is present at the input.

To increase the data transmission rate, but not the transmitted bandwidth, low-modulation-index FSK schemes must be employed. Fast-frequency shift keying (FFSK) [6] and minimum shift keying (MSK) [7] are two such bandwidth conservation techniques that have been developed. MSK is continuous phase FSK with a modulation index equal to 0.5. This choice of modulation index is significant because it results in an accumulated phase change of 90 deg over a single bit period and over each bit period the phase either progresses or regresses by this amount. Thus, the waveform of MSK exhibits phase continuity and since there are no abrupt phase changes as in QPSK, the waveform can be amplified by class C amplifiers without distortion. FFSK is similar to MSK except that the data input at the modulator is first differentially encoded. However, the MSK and FFSK-modulated signals still suffer from poor adjacent-channel sideband components due to their FSK-type spectrum, which makes them less suitable for digital radio systems where good spectral efficiency is required.

The spectral efficiency of an MSK system can be improved by prefiltering the binary data stream before it modulates the carrier. A Gaussian-shaped prefilter having the envelope shape

$$H(w) = \exp\left[-0.45 \cdot \left(\frac{w}{w_c}\right)\right] \qquad (8.5)$$

can be added to an MSK signal to produce a Gaussian MSK (GMSK) signal [8]. The bandwidth of the bit-shaping filter is usually defined in terms of the bandwidth-time product (BT). If $BT > 1$, then the waveform is essentially an MSK waveform. However, if $BT < 1$, intersymbol interference results and this effect is used to provide improved performance in the presence of noise, but only at the expense of increased complexity in the demodulator. BT values of 0.5 and 0.3 have been chosen for the Pan-European Cellular Mobile Radio System (commonly known as

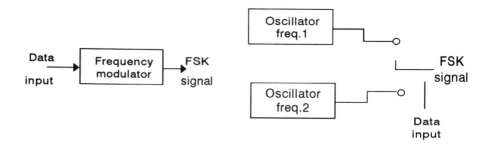

Figure 8.5 Generation of FSK.

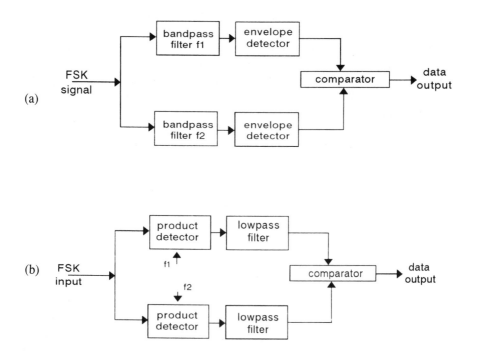

Figure 8.6 Noncoherent (a) and synchronous (b) detection of FSK.

the GSM system) and the Digital European Cordless Telecommunications (DECT) System, respectively.

An alternative approach to the Gaussian filtering is to correlatively code the data stream so that smaller phase changes occur. Tamed frequency modulation (TFM) [9] and generalized TFM (GTFM) [10, 11] are examples of such systems. TFM operates with the code rule which ensures that a change of phase by $\pi/2$ only takes place if three succeeding bits have the same polarity; no phase change takes place if the three bits are of alternating polarity and a phase change of $\pi/4$ occurs for the remaining bit configurations. The result is a much more reduced change in phase with a correspondingly much-improved spectral performance.

The modulator circuits necessary to generate these waveforms are similar and use one or two balanced modulators, and this has led to the concept of a generalized modulator structure [12] as shown in Figure 8.7. This is a quadrature-type modulator and the appropriate phase patterns for the particular waveform being generated are stored in the two EPROMS.

The demodulation of these signals requires a similar but reverse-ordered processing. In coherent demodulation circuits, on one hand, carrier synchronization techniques must also be included. On the other hand, noncoherent demodulation avoids this additional complexity but with some performance penalty.

Many digital communications systems have used constantenvelope modulation schemes such as TFM and GMSK because these schemes allow the use of less costly class C power amplifiers at cell sites. However, better spectral efficiency can be obtained by the use of linear modulation.

8.2.5 The Spectral Characteristics of the Modulation Schemes

For a given channel bandwidth, the power spectral density will affect the system efficiency and the narrower the modulation bandwidth the better the system efficiency, both in terms of interference and cost. The curves of the power spectral density versus normalized frequency fT_b are shown in Figure 8.8. The lowpass equivalent-power spectral density of PSK varies as $(\sin x)/x$. Thus if the input digital bit rate is f_b, the first zero point of the main lobe is equal to f_b. About 92.5% of the energy is contained in the main lobe and the amplitude of the spectrum decreases as $(f/f_b)^{-2}$. In PSK, only one of two possible signals can be transmitted during each signalling interval, but in QPSK one of four possible signals is transmitted so that QPSK has a bandwidth efficiency twice that of PSK.

With MSK, the first zero point of the power spectral density is at $0.75f_b$. The energy in the main lobe is about 95% and the amplitude of the spectrum decreases as $(f/f_b)^{-4}$. Thus, MSK has a narrower main lobe and much lower side lobes than PSK and in comparison to QPSK, MSK has lower side lobes but a wider main

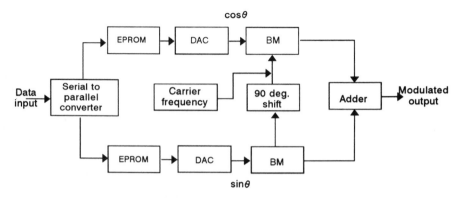

Figure 8.7 Universal I/O vector modulation.

lobe. Therefore in terms of spectral efficiency, there is a trade-off between QPSK and MSK that depends on the particular application [13].

The spectrum width of GMSK depends on the normalized 3-dB bandwidth of the lowpass Gaussian filter BT_b. When BT_b is reduced, the spectrum becomes narrower. If BT_b is equal to 0.2, the spectrum characteristic becomes similar to that of TFM, which has a narrower main lobe than MSK and no side lobe. If BT_b is infinite, the power-spectrum density is the same as that of MSK.

8.2.6 Error Rates

The function of a receiver in a data transmission system is to distinguish between the transmitted signals in the presence of noise and therefore a most important characteristic is the probability that an error will be made in such a determination. The probability of error in a system will depend on a number of factors, including type of modulation used, the type of detection employed, and the environment or channel. The two significant channel characteristics that need to be taken into consideration are multipath fading and multipath delay. For PSK, QPSK, and MSK, the probability of error P_e in the absence of multipath effects is

$$P_e = \frac{1}{2} \cdot erfc(E_b/n)^{1/2} \tag{8.6}$$

while TFM suffers a 1-dB degradation with respect to this performance assuming

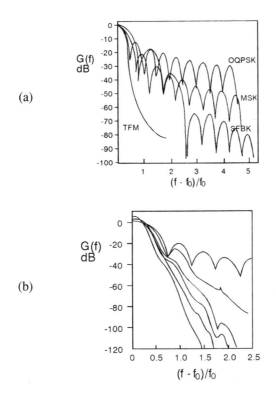

Figure 8.8 Power spectral characteristics of the modulation schemes: (a) PSD versus normalized frequency, (b) PSD of GMSK versus normalized frequency.

ideal clock recovery. A more detailed comparison of the different digital modulation techniques is given in [14].

8.2.7 Conclusions

Many factors such as delay, fading, and transmission bit rate affect the performance of the modulation scheme. When the transmitter and receiver are stationary and the signal path is constant, all the modulation schemes mentioned are suitable and PSK is preferred because of its lower cost and complexity, but if bandwidth consideration is important QPSK would be better. In the Rayleigh fading and delay environment GMSK, which has a lower error rate and comparably narrower bandwidth, would be more suitable. However, even with these modulation schemes, the maximum allowed bit rate is relatively low. To increase this, additional signal processing, such as channel coding, equalization, and diversity is required.

8.3 SPREAD-SPECTRUM MODULATION

Spread-spectrum systems [15–17] are wideband systems in which the bandwidth of the transmitted signal is much greater than that of the message signal. Different users are distinguished by means of different pseudorandom signature sequences or codes. These codes, or spreading signals, are combined with the message signal to produce a signal that occupies a bandwidth that is much greater than that of the data rate. Thus, the spread-spectrum signal occupies a bandwidth in excess of that necessary to send the information and produces a waveform that interferes in a barely noticeable way with any other signal operating in the same frequency band. There are two fundamental types of spread spectrum: direct sequence and frequency hopping.

8.3.1 Direct Sequence Spread Spectrum

In direct sequence (DS) spread spectrum an already modulated carrier is modulated a second time by a wideband spreading signal for purposes of bandwidth spreading. The spreading waveform can be generated by using a pseudonoise (PN) code generator. The PN sequence can be generated readily by using shift registers and modulo-2 adders. Although the sequence is generated in a deterministic manner, it can be considered to be random since the sequence length before repetition is usually extremely long, so that there is no correlation at all between the value of a particular bit and the value of any other bits. The PN sequence clock, which is called the chip rate, is much faster than the information signal clock. The transmitted sequence will be at the chip rate, but will contain the information of the first signal as well.

At the receiver, the incoming DS spread spectrum signal is first multiplied by an identical PN sequence to that used at the transmitter and then by the carrier waveform. The resulting waveform is integrated over the bit duration and the output of the integrator is sampled. Thus, at the receiver it is necessary to regenerate the carrier frequency and also to regenerate the PN sequence in synchronism with the transmitter spreading code. A block diagram of a DS spread spectrum communication system is shown in Figure 8.9.

One of the major disadvantages of DS spread spectrum systems is the so-called near-far effect, which can arise when the receiver signal-power level from different users varies considerably. Since each user utilizes the same bandwidth at the same time, it is possible for the power of a nearby transmitter to effectively swamp the signal from other users. This can be overcome in a star topology [18] by the base station continually transmitting some form of continuous tone to each user in the cell, who monitors the received signal level. The users then work out the characteristics of the channel and vary their transmitting power level inversely proportional to the level of the received signal. The down link is controlled by the base station, which can monitor the channel and adjust power levels accordingly. However, adaptive power control adds further to the cost and complexity of spread-spectrum systems.

8.3.2 Frequency Hopped Spread Spectrum

Frequency hopped (FH) spread spectrum is a FM or FSK technique. In FH spread spectrum the spectral density of the data-modulated carrier is widened by changing

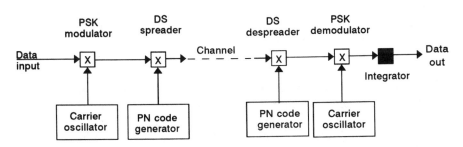

Figure 8.9 DS spread spectrum system.

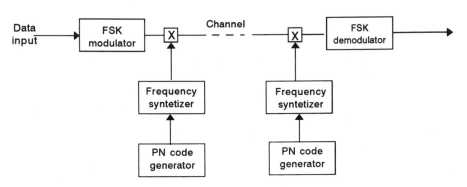

Figure 8.10 FH spread spectrum system.

the frequency of the carrier at a rate called the hopping rate. An FH spread-spectrum communications system is shown in Figure 8.10.

At the transmitter, the data is first modulated onto a carrier using conventional FSK (other modulation techniques, such as PSK, could also be used). The frequency hopping is then accomplished by using a mixer circuit where the mixing frequency is provided by the output of a frequency synthesizer whose frequency is changed by the PN spreading code.

The frequency synthesizer in the receiver is hopped in synchronism with that at the transmitter. This despreads the FH spread-spectrum signal and the original data can be recovered by use of a conventional FSK demodulator.

FH spread spectrum systems are excellent for combatting frequency-selective fading, which is a characteristic of the indoor channel. The FH system hops through the whole frequency range very rapidly, never spending much time in any one fade, so that the average fade is much less. This in turn allows the fading margin to be reduced. There is also no near-far problem with this spread spectrum technique because interfering sources will not be at the same frequency as the desired source.

FH spread spectrum systems can be divided into two broad categories depending on the number of frequency hops that are transmitted per bit of information. If the system changes frequency every bit then it is described as a fast frequency-hopping system, while if multiple bits are transmitted per frequency hop, it is described as a slow frequency-hopping system. The frequency-hopping rate can be selected by taking into consideration the fading characteristics of the channel. FH spread spectrum systems are more complex and costly to implement than direct sequence systems. The major cost is in the frequency synthesizers, which become more expensive and difficult to implement as the hop rate increases.

Another interesting performance characteristic of spread spectrum systems is the processing gain. This is related to the ratio of the channel bandwidth to the information bandwidth and leads to an improvement of the same order in the system signal-to-interference ratio.

8.4 CHANNEL ACCESS SCHEMES

A multiuser environment requires some form of multiple access to provide transmission facilities for individual users. The options for multiple access fall into three broad categories [9]: frequency division multiple access (FDMA), time division multiple access (TDMA), and code division multiple access (CDMA).

8.4.1 Frequency Division Multiple Access

In FDMA, the total spectrum assignment is divided into channels in the frequency domain. The total number of channels available at any one time are assigned on demand on a first-come, first-served basis to users initiating a call or to whom an incoming call is directed and once the channel has been assigned it is used exclusively for that communication. FDMA channels are relatively narrowband and the trend is towards even narrower bandwidths. FDMA systems tend to be less complex than either TDMA or CDMA systems and because the transmission is continuous the FDMA system will maintain synchronization with fewer bits devoted to overheads than TDMA or CDMA systems.

The disadvantages of the FDMA system are that it requires considerably more equipment at the base station to handle a given number of subscribers because of the single-carrier per channel design and because the capacity of the system can only be increased by increasing the total bandwidth available to the system. Because of the continuous nature of the transmission, FDMA systems require duplexers to prevent the mobile's transmitter interfering with the mobile's receiver. Also, the continuous nature of the transmission makes the implementation of handoff to another channel much more difficult than in a TDMA system where the change can be made during an idle time slot. Existing analogue cellular systems use FDMA [20,21].

8.4.2 Time Division Multiple Access

With TDMA [22,23], the channels are multiplexed by time division so that each channel accesses the full bandwidth for a short time slot. The total number of simultaneous users is limited by the number of time slots that are available, and users only use the channel during specific time slots. The total bandwidth for the TDMA system is determined by the number of channels to be multiplexed and by the choice of modulation technique to be used. A guard band is required between the time slots to ease synchronization requirements, to allow for timing drift between individual terminals, and to allow for different propagation delays between terminals and the base station. Because of the need for guard bands and the need for the receiver to reacquire synchronization on each burst, TDMA systems usually require much more overhead than FDMA systems and thus TDMA systems tend to be more complex. In addition, the higher channel rates proposed for some TDMA systems can create a much more severe intersymbol interference problem than FDMA systems.

The major advantage of TDMA systems over FDMA is the reduced cost of central-site equipment, which arises because each radio channel is effectively shared by a much larger number of subscribers. Other advantages include the cost saving because of the elimination of the duplexer circuitry and the more efficient handover procedure possible because of the on-off nature of the transmission. In addition, TDMA is more flexible, can accommodate different rate transmitting sources, and is more open to technological change.

8.4.3 Code Division Multiple Access

CDMA is the characteristic form of multiple access that is used for spread spectrum systems. In these systems each unit is assigned a unique randomized code sequence, different from all other users. Spread spectrum systems utilize a single wideband carrier, and thus in CDMA systems a large number of users can transmit simultaneously, resulting in the bandwidth being very wide when compared to either TDMA or FDMA systems. The bandwidth is also much wider than the likely coherence bandwidth of any multipath fading and, hence, tends to average out the frequency selective and signal-strength variations in the fading signal. Thus, CDMA is considered very appropriate for communications systems such as radio-based networks, which are likely to be subject to high interference levels and to multipath fading. Because of the very high bit rates, the symbol times are very short and can be less than the average delay spread so that intensive equalization is necessary.

In comparing the different multiple access schemes a number of factors, such as complexity and cost, network requirements, interference susceptibility, and spectrum efficiency, must be taken into consideration. CDMA mobile units are of comparable complexity to TDMA units and therefore costs should be similar, though in FH spread spectrum systems the requirement for frequency hopping is likely to make CDMA units more costly initially. On the other hand, because of the very large number of subscribers sharing the cost of the base-station channel equipment, the overall system-cost per subscriber will be low. In theory the capacity of the three systems using the same bandwidth is nearly equal. However, if factors such as power control, voice activation, and spatial isolation are taken into consideration [24], the capacity of the CDMA system is significantly higher than either FDMA or TDMA.

Since no one solution is optimal, it is not surprising that several hybrid solutions are being considered. For example, the DECT system uses time division duplex (TDD) transmission with a combination of 12 time-slots per carrier TDMA and 10 carriers per 20 MHZ of spectrum FDMA. Thus, DECT is a TDD/TDMA/FDMA system. Various other such combinations of FDMA/TDMA/CDMA and FDD/TDD are also possible.

8.5 EQUALIZATION

In a communications system the characteristics of the channel may vary as a function of a particular connection or line used as well as possible variations with time. In such systems it is advantageous to include a filter that can be adjusted to compensate for the nonideal characteristics of the channel. Such filters are called equalizers. In radio communications, the transmission is affected by the time-varying multipath propagation characteristics of the radio channel. To compensate for these time-varying characteristics, it is necessary to use time-varying or adaptive equalization. This can be accomplished by transmitting a known bit sequence along with the information bits and the receiver can then derive the transfer characteristic of the channel at the time of transmission by comparing the actual received signal with the known bit sequence. The receiver then uses an appropriate processing algorithm to correct the errors that occur in the subsequent information bits.

The time dispersion of the transmitted signal caused by multipath propagation causes intersymbol interference (ISI), and by compensating for ISI adaptive equalization will increase the data rate of the system. There are three main groups into which adaptive equalizers can be subdivided: linear transversal equalizers, non-linear or differential feedback equalizers, and maximum likelihood sequence estimators.

8.5.1 Linear Transversal Equalizers

The transversal filter consists of a tapped delay line, tap weight multipliers and a summer. If the filter has only feed forward taps, the transfer function is a polynomial in z^{-1} (poles are at $z = 0$) and it is a nonrecursive, finite impulse response (FIR) filter, as shown in Figure 8.11.

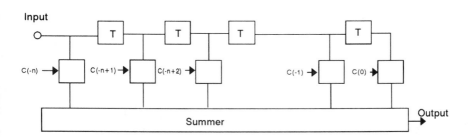

Figure 8.11 Linear transversal equalizer.

The delay T or symbol spacing is sensitive to the timing phase or delay of the channel. The linear equalizer tends to invert the channel frequency response and therefore at frequencies severely attenuated by the channel, white noise at the input to the receiver will be considerably amplified. For T-spaced equalizers, the channel is sampled at rate $1/T$ and the equalizer is sensitive to distortion near the frequency $f = \frac{1}{2}T$ caused by delay distortion or poor choice of sample timing phase. If the delay taps are spaced at intervals of τ, which is less than the symbol rate (i.e., $\tau < T$), a better performance will result. Such equalizers are known as fractionally spaced equalizers (FSE) [25].

8.5.2 Decision Feedback Equalizers

The most common nonlinear equalizer is the decision feedback equalizer (DFE) [26]. A DFE consists of two parts, a feedforward section and a feedback section, as shown in Figure 8.12. The feedforward section compensates for delay distortion whereas the feedback section compensates for slope distortion, which relates to the data rate of change of the mean squared error. The received (known) sequence is fed into the feedforward section, which is a linear transversal filter with the tap spacing equal to one symbol period. Decisions about the received symbols are then made by the feedforward section and fed into the feedback section, whose function is to remove any ISI from the present estimate caused by these symbols. The advantage of DFE is less noise enhancement than linear equalizers for severely distorted channels. However, when a channel is severely distorted, a DFE is not the optimum system for equalization since the DFE essentially discards all trailing echoes by merely subtracting their effects from the signal. After the prefiltering stage the DFE cancels the ISI from future signals, resulting in the loss of some signal power.

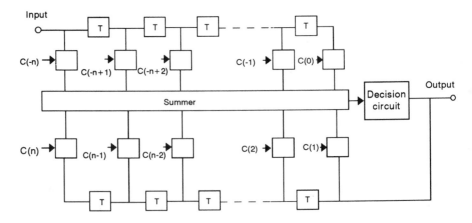

Figure 8.12 Decision feedback equalizer.

8.5.3 Maximum Likelihood Sequence Estimators

A maximum likelihood sequence estimator (MLSE) receives the complete data signal and correlates this complete signal with all possible transmitted symbols. The symbol that gives the largest correlation is then detected as the transmitted symbol. Unfortunately, the amount of computation for such a system increases exponentially with the length of the symbol message [27]. The choice among the various algorithms that have been proposed for adaptive equalizers depends on the requirements of the specific application, particularly convergence rate, distortion, and complexity. The principal challenge in designing equalization algorithms for mobile-radio applications is in the requirement for continuous adjustment to very rapid and large changes in the channel characteristics. Comparative performance results for a range of equalizer algorithms are given in [28].

8.6 CODING

Factors such as delay, fading, and multipath propagation affect the performance of a mobile-communication system and can result in a high bit-error rate. Equalization and diversity can give some improvement in performance, but further improvement may be necessary. Error control falls into two basic categories: error detection and retransmission, and forward error correction. The first method involves the use of a code to detect errors and to automatically request a repeat (ARQ) of any block containing errors. An ARQ system requires a feedback channel to the transmitter and suitable buffering. Also, the decoding delay is variable and dependant on the loop delay of the system. Forward error correction (FEC) [29] is achieved by means of channel-coding systems, which are based on the principle that the bit sequence can be encoded prior to transmission in such a manner that errors can be not only detected but also corrected by the receiving circuitry (see Fig. 8.13). Channel coding provides improved performance because of redundancy and noise-averaging affects. There are two broad categories of channel codes: block codes and convolutional codes [30,31].

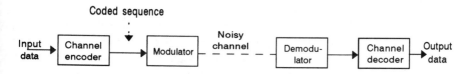

Figure 8.13 Channel-coding system.

8.6.1 Block Coding

A block code is a mapping of k input binary symbols into a larger number n output binary symbols. Hence, the block coder is a memoryless device. Since $n > k$, the code can be selected to provide redundancy, such as parity bits, which are calculated from the information bits and used by the decoder to provide some error detection and correction. The codes are denoted (n, k) and the code rate R is defined as $R = k/n$. The ability of a particular block code to detect and correct errors is dependent on the minimum number of positions in which any two n-digit code words in a system differ. This number is called the minimum or Hamming distance of the code and is denoted by d. At the detector the received code word can be checked for errors. If there are s errors in the received code word, then provided $d < s + 1$, it will be possible to detect with certainty that the received word is invalid, so that errors must have occurred. If there are t errors in the received word, then provided $d \geq 2t + 1$, it will be possible to both detect and correct the errors.

Decoding is a much more complex process than encoding. The most general optimum decoding algorithm computes the Hamming distance between the received code word and all possible code words and selects the word with the smallest distance as the one most likely to have been transmitted. Even for fairly moderate block lengths this is a very complex algorithm. The use of linear codes helps to reduce the complexity of the decoder, but on its own is not enough to make block codes practical, so additional structure, as is provided by cyclic codes, is required. Cyclic codes are block codes such that another code word can be obtained by taking any one code word, shifting the bits to the right and placing the dropped-off bits on the left. These types of codes have the advantage of being easily encoded from the message source by the use of inexpensive linear-shift registers with feedback. This structure also allows these codes to be easily detected by practical and efficient decoders. Examples of cyclic and related codes are Reed Solomon (RS) codes, Golay codes, and Bose-Chaudhuri-Hocquenghem (BCH) codes [29,30,31].

8.6.2 Convolutional Codes

In contrast to block codes, the parity-check information in convolutional codes is distributed over a span of message symbols, called the constraint span of the code. In this way long streams of message bits can be encoded continuously without the necessity of grouping them into blocks. This is accomplished by using shift registers whose outputs are combined in a preset manner to give certain constraints within the encoded bit stream. Thus, a convolutional code of constraint span K can be generated by combining the outputs of K shift registers in a predetermined manner in τ modulo-2 adders. At each clock time the outputs of the adders are sampled and thus τ output symbols are generated for each input symbol, giving a code rate of $1/\tau$. A block diagram of a convolutional encoder is shown in Figure 8.14.

A convolutionally encoded signal is decoded by matching the encoded-received data to the corresponding bit pattern in a code tree, which is a representation of the possible output codes that can be obtained from a particular convolutional encoder. In sequential decoding, the decoder follows a path through the code tree and, depending on the branch taken, a decision on the transmitted data is made. If noise is present in the channel, some of the received bits might be in error and then the paths will not match exactly. In this case a match is found by choosing a path that will minimize the Hamming distance between the selected-path sequence and the received-encoded sequence.

An optimum decoding algorithm, called Viterbi decoding, uses a similar procedure. It examines the possible paths and selects the best ones based on some conditional probabilities. The optimum Viterbi procedure uses so-called soft decisions. In hard-decision decoding the demodulator output is a binary sequence, which means that the demodulator is making hard decisions, whereas in soft decision decoding the demodulator passes on three or more level versions of its output to the decoder, which improves the decoder accuracy. Soft-decision decoding increases the receiver complexity, but can substantially improve the performance of a communications system.

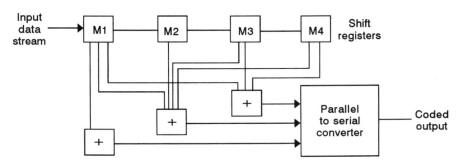

Figure 8.14 Convolutional encoder.

8.6.3 Code Interleaving and Concatenated Codes

The improvement in performance of a digital-communication system that can be achieved by the use of coding is measured in terms of the coding gain. The coding gain is defined as the reduction in SNR that is achieved when coding is used when compared to the SNR required for the uncoded case to achieve the same level of bit error rate. Most codes are designed to correct random errors, which assume that the channel is memoryless. However, a fading channel, especially where the fading is slow compared to the symbol period, has memory. This is particularly true for the indoor channel, which suffers from slow (lognormal) fading in addition to fast (Rayleigh) fading. In this case, even though the average bit-error rate is small, these codes are not effective in correcting errors because the errors are clustered. In this case bursts of errors will occur at the decoder output because the noise bursts are larger than the redundancy time of the code. A primary technique that is effective in overcoming burst errors is interleaving. Interleaving can be used with both block and convolutional coding systems. At the transmitter the coded data goes through a process of interleaving by shuffling the coded bits over a time span of several block or constraint lengths [32]. The time span is chosen to be several times the duration of the noise burst. At the receiving end the data bits are deinterleaved to produce coded data with isolated or random errors, which can then be corrected by passing the data stream through the decoder. This process produces an almost error-free output even when burst errors occur during the transmission.

While the use of interleaving is an effective way of dealing with error bursts, it is not particularly effective in dealing with random errors that affect only a single bit or only a small number of consecutive bits. Because of signal fading and delays due to multipath transmission in the indoor wireless channel, errors occur neither independently at random nor in well defined bursts, but in a mixed manner. One technique to combat these mixed errors is to cascade coding, which is effective in combatting error bursts and coding that reduces random errors. Such cascading of codes is called concatenation [33] and this has been shown to be an effective technique in reducing errors over fading dispersive channels [34].

Forward error-correcting techniques can significantly improve performance, but their use in wireless network systems will be constrained by both the bandwidth limitations and the complexity involved in implementing the more advanced algorithms.

8.7 SYNCHRONIZATION

The coherent demodulation of data signals, which have been transmitted over a radio channel, is based on the assumption that exact replicas of the transmitted waveforms can be reproduced at the receiver. This is only possible if the carrier frequency and phase and the symbol timing can be precisely extracted from the received signal. Any error in the phase of these reference signals causes a degradation in the performance of the detector.

8.7.1 Carrier Recovery

There are a variety of different techniques for recovering accurately the carrier frequency. One of the simplest carrier recovery circuits is the squaring loop shown in Figure 8.2. In this circuit the received signal is passed through a square law device and then bandpass filtered, which reduces or eliminates the phase effects of the modulating signal. The reference carrier is subsequently obtained by frequency division by a factor of two. For M-ary PSK, the carrier can be recovered by an Mth power method, which is similar to the squaring loop. In this method, the received signal, after initially passing through a bandpass filter to remove the noise outside the band required for the signal itself, is passed through a network whose output is its input raised to the Mth power. The output of the Mth power circuit is passed through a narrowband filter to isolate the signal at Mf_c from the various other spectral components that will also be generated by the circuit. A divide-by-M circuit then yields the desired carrier to frequency f_c.

An alternative circuit for carrier recovery is the Costas loop [35]. This differs from the squaring loop in how it eliminates the modulating signal effects and how it generates the input to the tracking loop. The circuit involves two phase locked loops (PLLs) employing a common VCO and loop filter, as shown in Figure 8.15. Assume initially that the VCO is locked to the carrier frequency, but with a phase error Θ_e. If the received signal is $m(t) \cos(W_c t + \Theta)$, then the voltages $V_1(t)$ and $V_2(t)$ at the outputs of the lowpass filters are

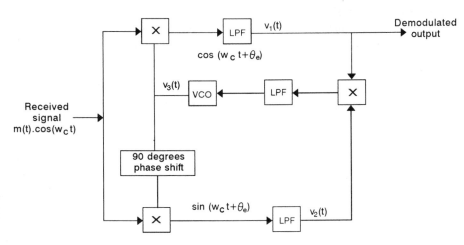

Figure 8.15 Cost as loop for carrier recovery.

$$v_1(t) = \frac{m(t)}{2} \cdot \cos \Theta_e$$

$$v_2(t) = \frac{m(t)}{2} \cdot \sin \Theta_e$$ (8.7)

and after multiplication and filtering the signal

$$v_3 = \frac{m^2(t)}{8} \cdot \sin 2\Theta_e$$ (8.8)

is passed to the VCO.

Since Θ_e is small, the amplitude of $V_1(t)$ is relatively large compared to $V_2(t)$, and since $V_1(t)$ is proportional to $m(t)$, is it also the required output.

8.7.2 Bit Synchronization

A bit synchronizer is a circuit that regenerates at the receiver a clock waveform which is synchronous with the original clock waveform at the transmitter. Bit synchronizers can be extremely complex when the clock waveform must be extracted from a corrupted received signal, and the complexity of the circuit will also depend on the synchronizing properties of the line code.

Clock recovery is usually accomplished by operations on the transmitted data sequence. One method of achieving this relies on the threshold crossings of the data sequence and a very stable crystal oscillator, which counts at a rate several orders of magnitude higher than the bit rate and effectively counts down from one sampling time to the next. Another common method for clock recovery uses a squaring loop, as shown in Figure 8.16. The received bit stream is prefiltered and

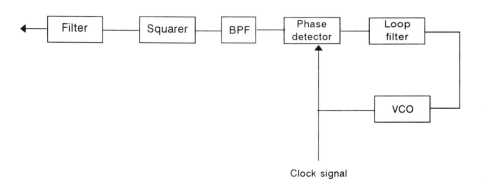

Figure 8.16 Clock recovery circuit.

then passed to a squaring circuit, which eliminates the effects of the particular data sequence. The squarer output is then sent to a bandpass filter tuned to the symbol rate and the tracking loop extracts the clock signal from the filter output.

Various other clock recovery circuits exist and for further information on synchronization the reader is referred to a special issue of the *IEEE Transactions on Communications* [32,36].

8.8 SUMMARY

The basic signal-processing techniques and technology related to transmitting data over mobile RF channels have been presented in this chapter. The emphasis has been placed on those techniques that are considered to be most appropriate to wireless LAN systems. In choosing the modulation scheme for a wireless environment, the emphasis is usually on trying to provide high spectral efficiency. Both narrowband modulation schemes and spread spectrum techniques, which are emerging as strong contenders for wireless network systems, were discussed. The three broad categories of multiple access are FDMA, TDMA, and CDMA, and while TDMA is currently the preferred scheme, the emergence of spread spectrum systems will result in CDMA becoming more widespread in the future.

Multipath fading and propagation delays affect the performance of radio-based communications systems. Equalization will give some improvement in performance, and error-control techniques can be used to give further improvement. Block and convolutional coding schemes are discussed. However, in the fading channel additional techniques such as interleaving and concatenation may be necessary to overcome the various propagation induced errors that can occur. Finally, the problem of receiver synchronization, which is essential for accurate detection of the data stream, is covered. While sophisticated modulation, equalization, coding, and synchronization techniques will enhance the quality of transmission, their use in radio-based network systems will tend to be constrained by both bandwidth limitations and the increased cost and complexity of implementing the more advanced systems.

REFERENCES

[1] Suzuki, H., and K. Hirade. "Spectrum efficiency of *M*-ary PSK Land Mobile Radio," *IEEE Trans. on Communications*, Vol. 30, July 1982, pp. 1803–1805.

[2] Chuang, J.C.I. "The Effects of Time Delay Spread on Portable Radio Communications Channels with Digital Modulation," *IEEE Journal on Selected Areas in Communications*, Vol. 5, June 1987, pp. 879–889.

[3] Chuang, J.C.I. "The Effects of Time Delay Spread on QAM with Nonlinearly Switched Filters in a Portable Land Mobile Radio Communication Channel," *Proc. Intl. Conf. on Digital Land Mobile Radio Communications*, Venice, June 30–July 3, 1987, pp. 104–113.

[4] Akaiwa, Y. and Y. Nagota. "Highly Efficient Digital Mobile Communications with a Linear Modulation Method," *IEEE Journal on Selected Areas in Communications*, Vol. 5, June 1987, pp. 890–895.

[5] Gronemeyer, S. A., and A. L. McBride. "MSK and Offset QPSK Modulation," *IEEE Trans. on Communications*, Vol. 24, August 1976, pp. 809–820.

[6] Holbeche, R. J. (ed.). *Land Mobile Radio Systems*, London: Peter Peregrinus, 1985.

[7] Pasupathy, S. "Minimum Shift Keying: a spectrally efficient modulation," *IEEE Communications Magazine*, July 1979, pp. 14–22.

[8] Murota, K., and K. Hirade. "GMSK Modulation for Digital Mobile Radio Telephony," *IEEE Trans. on Communications*, Vol. 29, July 1981, pp. 1044–1050.

[9] de Jager, F., and C. B. Dekker. "Tamed Frequency Modulation, a Novel Method to achieve Spectrum economy in Digital Transmission," *IEEE Trans. on Selected Areas of Communications*, Vol. 26, May 1987, pp. 534–542.

[10] Chung, K. S., and L. W. Zegers. "Generalized Tamed Frequency Modulation," *Philips Journal of Research*, Vol. 37, 1982, pp. 165–177.

[11] Burkley, C. J., C. O'Donoghue, and M. O'Droma. "The Performance of GTFM in a Frequency Selective Rayleigh Fading Channel," *39th IEEE VTC*, San Francisco, May 1989, pp. 878–883.

[12] Davarian, F., and J. T. Sumida. "A multipurpose digital modulation," *IEEE Communications Magazine*, February 1989, p. 36.

[13] Amoroso, F. "The Bandwidth of Digital Data Signals," *IEEE Communications Magazine*, Vol. 18, November 1980, pp. 13–24.

[14] Oetting, J. D. "A Comparison of Modulation Techniques for Digital Radio," *IEEE Trans. on Selected Areas of Communications*, Vol. 27, December 1979, p. 1757.

[15] Dixon, R. C. *Spread Spectrum Systems*, 2d ed., New York: John Wiley & Sons, 1984.

[16] Kavehrad, M., and P. McLane. "Spread Spectrum for Indoor Digital Radio," *IEEE Communications Magazine*, June 1987, pp. 32–40.

[17] Simon, M. K., J. K. Omura, R. A. Scholtz, and B. K. Levitt. *Spread Spectrum Communications*, Vols. 1–3, Rockville, MD: Computer Science Press, 1985.

[18] Kavehrad, M., and B. Ramamurthi. "Direct Sequence Spread Spectrum with DPSK Modulation and Diversity for Indoor Wireless Communications," *IEEE Trans. on Selected Areas of Communications*, Vol. 35, February 1987, pp. 224–236.

[19] Li, V.O.K. "Multiple Access Communications Networks," *IEEE Communications Magazine*, June 1987, pp. 41–48.

[20] Tarallo, J. and Zysman, G. I. "A Digital Narrowband Cellular System," *Proc. 37th IEEE VTC*, Tampa, 1987, pp. 279–280.

[21] Russell, J. "Digital Cellular Takes the FDMA Route," *Communications Systems Worldwide*, April 1988, pp. 44–48.

[22] Uddenfelt, J., and B. Persson. "A narrowband TDMA System for a New Generation Cellular Radio," *Proc. 37th IEEE VTC*, Tampa, June 1987, pp. 286–292.

[23] Bolgiano, D. R. "Spectrally Efficient Digital VHF Mobile System," *Proc. 38th IEEE VTC*, Philadelphia, June 1988, pp. 693–696.

[24] Gilhousen, K. S., I. M. Jacobs, R.Padovani, and L. A. Weaver. "Increased Capacity Using CDMA for Mobile Satellite Communications," *IEEE Trans. on Selected Areas in Communications*, Vol. 8, May 1990, pp. 503–514.

[25] Qureshi, S.U.H. "Adaptive Equalization," *IRE Proc.*, Vol. 73, September 1985, pp. 1349–1387.

[26] Befiore, C. A., and J. H. Park. "Decision Feedback Equalization," *IRE Proc.*, Vol. 67, August 1979, pp. 1143–1171.

[27] Lucky, R. W. "Techniques for Adaptive Equalization for Digital Communications," *BSTJ*, Vol. 44, April 1965, pp. 547–588.

[28] Hirsch, D., and W. J. Wolf. "A Simple Adaptive Equalizer for Efficient Data Transmission," *IEEE Trans. on Selected Areas of Communications*, Vol. 18, February 1970, pp. 5–12.

[29] Bhargara, V. K. "Forward Error Correcting Schemes for Digital Communications," *IEEE Communications Magazine*, January 1983, pp. 11–19.

[30] Lin, S., and D. J. Costello. *Error-Control Coding, Fundamentals and Applications*, Englewood Cliffs, NJ: Prentice-Hall, 1983.

[31] Clark, G. C., and J. T. Cain. *Error-Correction Coding for Digital Communications*, New York: Plenum Publishing Corp., 1981.

[32] Sklar, B. *Digital Communications*, Englewood Cliffs, NJ: Prentice Hall, 1988

[33] Forney, G. D. *Concatenated Codes*, Cambridge: MIT Press, 1966.

[34] Mokrani, K., and S. S. Soliman. "Concatenated Codes over Fading Dispersive Channels," *IEEE International Communications Conference*, Boston 1989, pp. 1378–1382.

[35] Costas, J. P. "Synchronous Communications," *IRE Proc.*, Vol. 44, December 1956, pp. 1713–1718.

[36] Gardner, F. M., and W. C. Lindsey (guest eds.). "Special Issue Synchronization," *IEEE Trans. on Selected Areas of Communications*, Vol. COM-28, August 1980.

Appendix 1
Safety and Regulations

Manuel J. Betancor, Francisco J. Gabiola,
Victor M. Melián, Victor A. Araña
E.T.S.I. Telecomunicación
Universidad de Las Palmas de Gran Canaria
Campus Universitario de Tafira, 35017 Las Palmas, Spain

A.1 INTRODUCTION

To draw up this appendix for infrared technology, we used studies and known standards for the safe use of lasers and fiber-optic systems because the radiation is similar to that produced by an infrared-emitting diode (IRED). With respect to the biological treatment of the eye and the damage that can be produced, the effect is caused by the wavelengths and optical output power used in optical communication systems. When the human body is illuminated by the radiation of the visible spectrum and the near infrared, the most sensitive tissues are the eye and the skin; the former is more vulnerable to damage than the latter. The American National Standards provides guidance for the safe use of lasers, laser systems, and fiber optics by defining control measures for each laser classification.

We will describe the effects of optical radiation on the eye and the maximum permissible exposure (MPE) to avoid damaging the tissues. The MPE values given in standards such as ANSI Z136.1 are below known hazardous values. Standards list MPE values for all wavelengths from 0.220 to 1000 mm and for exposure times from 10^{-9} to 3×10^{-4} sec. One table is for direct (or intrabeam) viewing and another is for viewing a diffuse reflection. Some authors suggest that classification tables for intrabeam and diffuse lasers must be modified in terms of power density for CW or thresholds (using experimental measurements) [1].

trum through the eye results in these wavelengths being retinal hazards because these light rays reach the retina and are absorbed there. The source size and image size are also important. Most of the LEDs or IRED used in wireless optical-communication systems emit wavelengths between 0.7 and 1 μm (i.e., 0.850, 0.930).

The concentration of light on the retina depends on the illumination of the object that is being viewed. For example, the sun is a collimated distant source of light and produces an image 160 μm in diameter on the retina: this is ten times the diameter of an image produced by the almost-parallel rays emitted by a laser.

Plots of MPE for several conditions of power and exposure time for commercial IREDs are presented. These are obtained from calculations of minimum safe distance for unresolved sources. The risk presented by these IREDs is studied using these calculations and standards for nonlaser sources as ACGIH TLV. Although the hazard of a GaAs IRED is basically no different from that of a laser diode of the same projected source size, the total radiant-power emission from IRED may be greater than that of a diode laser but the radiance is always less. Finally, a table of laser radiation standards for some countries is presented.

In closing we will study the safety standards concerning microwaves. The injury produced by microwaves depends on power levels and time exposure and it is noticed as heating. Although the potential hazards are biological, most of them are subjective: fatigue, headache, sleepiness, and so forth. Microwave exposure represents no hazards unless overheating. Heating depends on exposure time and field strength. We draw up the specific absorption rate (SAR) over tissues to get the minimum safe distance and power density to avoid damages when microwaves are used. We present plots of SAR, power density, and safe distance versus frequency for bands used in wireless local area networks.

A.2 EFFECTS OF OPTICAL RADIATION ON THE EYE

The optical radiation absorption properties of the eye vary significantly with wavelength. The absorption of electromagnetic radiation by the eye is shown in Figure A.1. Here we can see how the retina is not affected by the far ultraviolet and far infrared because both are absorbed in the outer layers of the eye.

The main damage to the retina is produced by wavelengths between 0.4 and 1.4 μm: visible and near infrared (this zone is commonly known as the retinal hazard region). Light and near-infrared radiation is sharply focused onto the retina. When an object is viewed directly, the light forms an image in the fovea. The typical result of a retinal injury is a blind spot, or *scotoma*, within the irradiated area.

The greatest risk to the eye tissues is presented by highly collimated sources, and this risk is augmented when the eye is relaxed (focused at infinity) because the focusing properties of the human eye create a very serious hazard when the

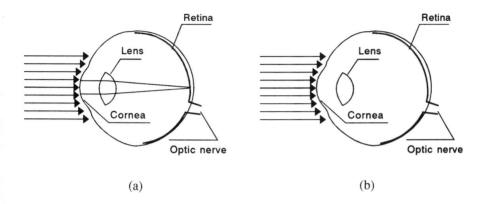

Figure A.1 Absorption and transmission of laser light by components of the ocular system: (a) visible and near infrared (0.4–1.4 μm); (b) for infrared (>1.4 μm).

eye is located within the beam (intrabeam viewing). Noncollimated beams or beams that have been deflected over diffused surfaces are much less risky [2]. This is the case of IREDs. If the eye is relaxed the retinal irradiance is increased by a factor of 100,000 times. Pupil size also influences the retinal irradiance.

Experiments helped show that there are no long-term cumulative effects of near-infrared laser radiation on visual function or retinal tissue [3]. At 860 nm, it required powers ranging from 44.3 to 27.7 mW entering the eye through the pupil for periods ranging from 10 to 1,000 sec to produce a minimal lesion in the macaque monkey retina. Image size is an important factor in determining temperature rise. For image diameter smaller than 500 μm, the power required to produce a lesion would be decreased by at least the ratio of the image diameters (e.g., a 100 μm diameter would requires 5 times less power to produce a lesion). Conversely, image diameters of >500 μm would require more power entering eye.

A.2.1 Injury to the Eye

The optical properties of the eye play an important role in determining retinal injury. Eye damage can be caused by thermal and photochemical processes and other mechanisms. The nature of the mechanism that causes the damage plays an important role in the definition of the degree of hazard, as does the fact of whether the effect is additional after a period of time or if it is possible that retarded consequences may appear. The potential biological effects of optical radiation are classified according to the wavelength themselves. At longer visible and near-infrared wavelengths retinal injury is produced via a thermal mechanism [1,4].

The near-infrared radiation may produce thermal burns in the retina. Part of this radiation is absorbed in the outer structure of the eye: the cornea, lens, the vitreous, and the aqueous (Fig. A.1). Although thermal lesions may be produced in these parts of the eye (particularly the lens), evidence supports the argument that the retina is still the most vulnerable ocular structure.

The radiation which corresponds to the visible region and the near-infrared is more significant given the number of sources (laser, light and infrared emitting diodes), whether coherent or otherwise, which emit within this zone of the spectrum. The focusing properties of the human eye are such that great risks are run when the eye is located within the beam. If the eye is relaxed (focused at infinity), the retinal irradiance is increased 100,000 times. For example, while the irradiance of a spotlight that is 20 μm in diameter is 100 mW/cm^2 at the cornea, the retinal irradiance is 10 kW/cm^2. If the focus is produced in the fovea (central area of the retina), it may produce a small blind spot (*scotoma*) in the eye of the person who is focussing at infinity within the beam.

The shorter wavelengths of visible spectrum are the most dangerous to the retina [4]. The longer the wavelength, the less the energy is absorbed in the retina, which means that retinal injury thresholds are typically 5 to 10 times greater for infrared wavelengths than for visible wavelengths. In order to take all of these effects into account, the maximum permitted exposure (MPE) has been defined.

In the case of diffuse reflection, a viewer is susceptible to injury if the source is close enough to be viewed as an extended source rather than a point source.

A.2.2 Pupil Size

The limiting aperture of the eye determines the amount of radiant energy entering the eye and therefore reaching the retina. The energy transmitted is proportional to the area of the pupil. The pupil varies its size from 2–7 μm when daylight turns into darkness. Therefore, during daytime the retina receives only 1/14th the light it receives in darkness. The angle subtended by the source also plays a role; thus a light source of a given luminance causes a different pupil size depending on viewing distance and the luminance of the surrounding field.

A.2.3 Eye Optics

The smallest image that can be resolved by the eye is about 10 μm. The angle subtended by an extended source defines the image size. Figure A.2 shows an eye looking at an extended source reasonably close to it. The size of the retinal image S is expressed in the following equation [5,6].

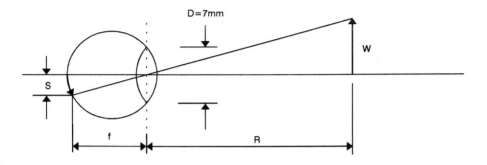

Figure A.2 Optic eye—an extended source is imaged at the retina.

$$S = \frac{Wf}{R} \tag{A.1}$$

where f is the focal length, R the viewing distance and W the dimension of the light source. The focal length f of the normal relaxed eye is about 1.7 cm. Thus, small objects produce a constant 10-μm diameter image for all ranges greater than those where S falls to 10 μm. In most practical cases with semiconductor lasers and LEDs, the image is unresolved and the power density decreases as R^2.

For small angles, the source area A_L and the image area A_R are always proportional to the ratio of the square of the viewing distance to the square of the eye's focal length. The retinal irradiance and source radiance are likewise proportional. Since solid angle of the source W_s is determined by its area, A_L and R, the radiance L of the source can be defined from corneal irradiance E_c:

$$L = \frac{E_c}{\Omega_s} = \frac{E_c R^2}{A_L} = \frac{E_c f^2}{A_R} \tag{A.2}$$

The total power that enters the eye through the area of the pupil A_C and reaches the retina (τ = transmission coefficient) is:

$$\Phi_R = E_R A_R = \tau E_c A_c = \tau E_c \left(\frac{\pi d_e^2}{4}\right) \tag{A.3}$$

where d_e is the pupil diameter.

Therefore, for small angles, the relation between the retinal irradiance and the source radiance is:

$$E_R = \frac{\pi d_e^2 \tau L}{4f^2} = 0.27 d_e^2 \tau L \tag{A.4}$$

The last equation is of great practical value since it permits the definition of a permissible radiance from a permissible retinal irradiance for any source of known radiance without concern for the viewing angle or viewing distance.

A.3 SAFETY STANDARDS FOR FIBER OPTICS

ANSI Z131.2 for laser safety in fiber optics defines four service groups of the optical-fiber communication systems based on the risk of injury. The accessible emission limit (AEL) for each service group is derived from the product of the maximum permissible irradiance (MPI) and the area of a specified limiting aperture. The fundamental method for determining the service group is to divide the power measured through the appropriate aperture, located at a specified distance, by the area of the limiting aperture and compare with MPI table. This standard has been written for optical communication systems utilizing laser diodes or light-emitting diodes (LD or LED) with average output powers of less than 500 mW and wavelengths between 0.4 μm and 1 μm. Following ANSI Z136.1, the emission from an optical fiber is usually divergent and therefore the risk for injury is different from that of a conventional laser with the same operating parameters (power and wavelength).

A.4 SAFETY STANDARDS FOR NONLASER SOURCES

We mentioned before that most of the studies used to draw up this appendix have focused on the effects of the radiation produced by a laser. This type of source is in the forefront of most safety standards [4,7]. However, several organizations, such as the American Conference of Governmental Industrial Hygienists(ACGIH), have drawn up and proposed limit values to avoid damage. ACGIH TLV limits the near-infrared radiation beyond 770 nm to 10 mW/cm^2 to avoid possible damage [7]. For long-exposure conditions and visible and near-infrared wavelengths, the limit condition for the open pupil (7 mm) is:

$$\sum_{700}^{1400} L_\lambda \Delta\lambda \; L(\text{hazard}) = \frac{0.6}{\alpha} \left(\frac{W}{cm^2 sr} \right) \tag{A.5}$$

A.5 MAXIMUM PERMISSIBLE EXPOSURE (MPE)

The maximum permissible exposure is given in terms of optical power density, that is *irradiance* (the units are mW/cm^2 or W/m^2) for continuous wave or optical energy density, and *radiant exposure* (the units are J/cm^2 or J/m^2), for pulsed beams. The MPI in terms of irradiance can be obtained by dividing the radiant exposure by time, t, in seconds; the MPI in terms of radiant exposure can be obtained by multiplying the irradiance by t. These values are conservative and have been found from experiments with retinal animal (i.e., macaque monkeys or rabbits). MPE is wavelength dependent and figures derived for 0.8 μm afford greater safety margins for 0.95 μm. Above 1.05 μm, the permitted energy level is 2.5 times that at 0.85 μm and the minimum safe distance is reduced by about a third.

As optical sources for wireless communications emit wavelengths in the range of near-infrared (0.7–1 μm) we present the MPE for these part of the spectrum. The MPE for exposure times from 0.18 μs to 1000 seconds for wavelengths between 0.7 and 1.05 μm can be obtained from the expression (ANSI MPE values for intrabeam viewing of a nearly point source):

$$MPE = 1.8C_A t^{3/4} \left(\frac{J}{cm^2}\right) \qquad (A.6)$$

where the constant C_A is:

$$C_A = 10^{2(\lambda - 0.7)} \qquad (A.7)$$

The MPE for exposure times from 10^3 to 3×10^4 can be obtained from:

$$MPE = 640C_A 10^{-6} \left(\frac{W}{cm^2}\right) \qquad (A.8)$$

The corneal maximum-permitted exposure versus wavelength for some exposure times are represented in Figure A.3. Here it can be seen that MPE increases with exposure time and wavelength. If the source is unresolved (10 μm in diameter), the maximum source output power to avoid retinal injury is:

$$\phi_{max} \leq (f/2)^2 MPE \qquad (A.9)$$

where f is the eye focal length. On the right-hand axis of Figure A.3 is the maximum source-optical power for open pupil ($f = 7$ mm).

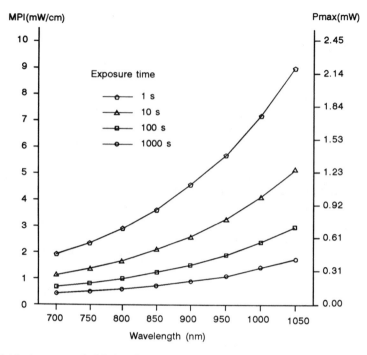

Figure A.3 Maximum permissible irradiance as a function of exposure time. Right-hand axis is the maximum source optical power for open pupil (7 mm).

A.6 MINIMUM SAFE DISTANCE

A solution to avoid eye injury when optical power is present is that the viewer does not see the source in a range limited for minimum safe-viewing distance. This distance takes into account the MPE, source optical output power, and numerical aperture of the source. The safe distance is derived for unresolved sources and is schematized in Figure A.4.

The next expression is used in optical-fiber communication systems. Radiation from an unterminated fiber must be assessed for hazard distance d_{min}, taking into consideration the power available at the nearest known assessment point, the distance, and the attenuation from that point. For multimode fiber, the minimum safe distance is:

$$d_{min} = \frac{1}{NA} \sqrt{\frac{\phi}{\pi MPE}} \qquad (A.10)$$

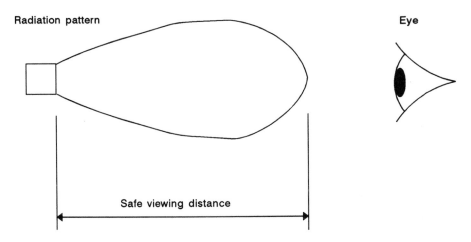

Radiation pattern

Eye

Safe viewing distance

Figure A.4 Diagram of safe viewing distance and IRED radiation pattern.

Figures A.5 and A.6 show the minimum safe distance versus IRED output power for several half-angle and for long exposure ($t = 2 \times 10^3$ sec). Both figures are for different wavelengths; Figure A.5 is for 0.85 μm and Figure A.6 is for 0.93 μm. Figures A.7 and A.8 show the minimum safe distance versus exposure time for an optical power of 10 mW and several wavelengths; Figure A.7 is for 0.85 μm and Figure A.8 is for 0.93 μm. It can be observed that figures for 0.85 μm afford greater safety margins than for 0.93 μm. The permitted energy for 0.93 μm is sometimes that at 0.85 μm. We have chosen these wavelengths because most commercial IREDs use them. The reason to use five half angle ranges is that available IREDs match them. Figure A.9 shows a bar diagram of half-angle for 376 commercial IREDs from 17 manufacturers. In all cases with a safe distance of 50 cm for optical power of 50 mW and highly directive IRED (5), eye damage is avoided.

A.7 RISKS FROM IRED

A GaAs source emitting at 900 nm is basically no different, from a hazard standpoint, than a GaAs laser diode of the same projected source size. The total radiant-power emission from an IRED may be greater than that of a diode laser, but the radiance is always less. By way of a general norm, infrared emitting diodes used in remote control or optical communication do not present risks. The emission

Dmin (cm)

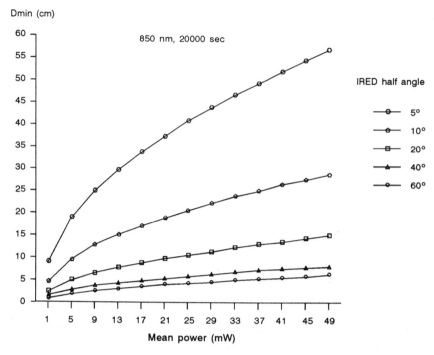

Figure A.5 Safe viewing distance as a function of optical power and IRED half angle. The exposure duration is 20×10^3 sec, the wavelength is 850 nm.

pattern of a diode laser or IRED is highly divergent, unlike that of a conventional laser, unless lenses are used. This means that the beam irradiance (risk for eye injury) decreases rapidly with distance from the output. Inadvertent viewing of fiber ends with the unaided eye at distances greater than 10 cm will not cause eye injury for power less than 10 dBm.

IREDs do not cause a hazardous retinal irradiance when one considers that bringing such an emitter closer than the near point of the eye (15 to 25 cm) can result in no greater concentration of infrared radiation since the image would increase with closer viewing distance.

Commercial IREDs have radiant power range 100 μW to 30 mW with half-angles from 5 deg to 60 deg. Some of the IREDs catalogued exhibit a diameter d of chip of about 0.25–0.5 mm. The radiance of these IREDs can be obtained using the projected source area and solid angle subtended for half-angle.

For most adults the near point of accommodation, minimum distance at which the eye can focus, is approximately 25 cm. For children and myopics, the near

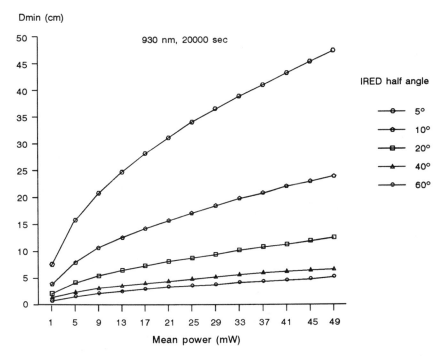

Dmin (cm)

930 nm, 20000 sec

IRED half angle

—⊝— 5°
—●— 10°
—☐— 20°
—▲— 40°
—⊘— 60°

Mean power (mW)

Figure A.6 Safe viewing distance as a function of optical power and IRED half angle. The exposure duration is 20×10^3 sec, the wavelength is 950 nm.

point may be as close as 10 cm. At this distance the angular subtense of the source α is:

$$\alpha = \frac{d(\text{cm})}{10(\text{cm})} \ (\text{rad}) \tag{A.11}$$

The ACGIH TLV criteria for a nonlaser IR source requires that the radiance should not exceed L_{hazard} for a distance of 10 cm:

$$L_{\text{hazard}} = \frac{0.6}{\alpha} \ \frac{W}{\text{cm}^2 sr} \tag{A.12}$$

Therefore, the IRED can be operated as a continuous wave (CW) or (more likely) as a train of pulses with the same peak intensity. Obviously, if the IRED

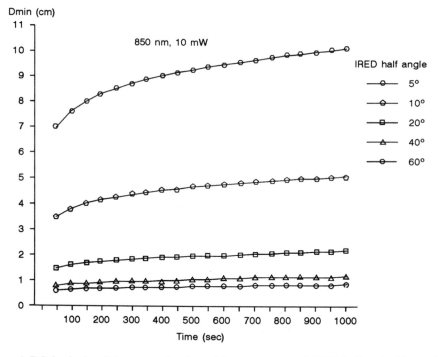

Figure A.7 Safe viewing distance as a function of time exposure and IRED half angle. The optical power is 10 mW, the wavelength is 850 nm.

is safe to view for the CW case, it is certainly safe if some of the total energy emitted each second were removed from the beam.

Table A.1 presents active area diameter, half-angle, and optical power for several situations where IRED radiance is below L_{hazard}. For an IRED array of 1 cm² ($l = 1.13$ cm), power would be 1.9W if the half-angle is greater than 20 deg; if half-angle is 15 deg, optical power must be less than 500 mW.

A.8 LASER RADIATION STANDARDS

In most countries, requirements are being developed based on the standard provided by the Technical Committee No. 76 of The International Electrotechnical Commission. In 1988, the European Organization for Electrotechnical Standardization (CENELEC) adopted a version of the IEC-825 standard as HD 482 S1. In 1992 that will be replaced with a European Norm, EN 60-825.

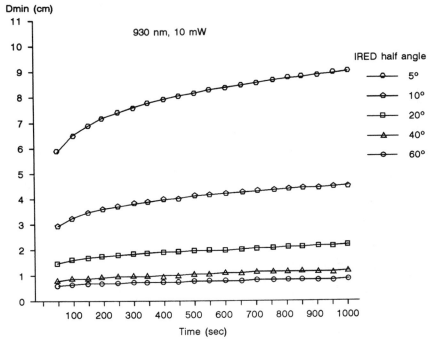

Figure A.8 Safe viewing distance as a function of time exposure and IRED half-angle. The optical power is 10 mW, the wavelength is 930 nm.

Many countries follow the user requirements found in the American National Standards Institute (ANSI) Standard for the Safe Use of Lasers. This document provides control measures, guidance for medical surveillance, and other helpful information for users. U.S. and IEC requirements differ primarily in labeling, interlocks, measurement criteria, and collateral (ancillary) radiation.

The ANSI Committee provides a standard on fiber optics for communications that covers both lasers and IRED sources for medical laser installations. Table A.2 presents countries and their laser standards.

A.9 RADIO MICROWAVE SAFETY

Microwave energy (300 MHz to 300 GHz) can produce biological effects or injury, depending on power levels and exposure durations. There is considerable agreement among scientists concerning the biological effects and potential health hazards

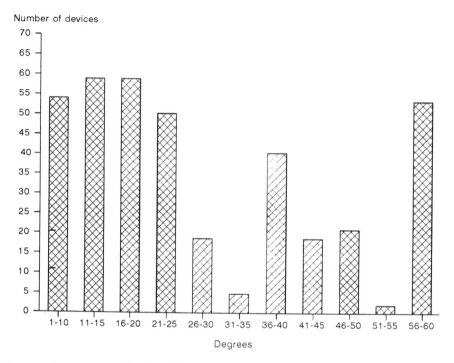

Figure A.9 Commercial IREDs classified by half-angle.

Table A.1
Active Area Diameter, Half-Angle, and Power for IREDs
or Arrays Without Hazard (ACGIH)

Diameter (cm)	Half-Angle (°)	Radiant Power (mW)
0.13	7.5	32
	15	125
0.5	5	56
	15	450
1.13	5	125
	15	1,000
	20	1,900

Table A.2
Laser Standards for Different Countries

Country	Standards
ANSI	Z136.1 (1986), Z136.2 (1988), Z136.3 (1988), Z136.4
Australia	AS2211—1981, AS2397—1980
Belgium	NBN C 79-700
Canada	Chapter 34, I.L.574, PC1977-3123, PC1977-3127
CENELEC	EN 60-825, HD 194, HD482-S1
China (People's Republic)	GB 7427-87
Denmark	IEC 825
Finland	SFS-IEC 825
France	NEC 74-311, NFC 43-801, NFC 74-312
Germany	DIN 56912, DIN 58126-P6, DIN 58215/52819, DIN/VDE 0835, DIN/VDE/0750, VBG 93 (1988), VDE 0836, VDE 0837
Holland	1978/6
IEC	1010-1, 1040, 601-2, 820, 825
Israel	1249,Part 1
Italy	CEI 76-1
Norway	NEK-HD 482 S1
Sweden	AFS 1981:9, SSI FS 1980:2, SSI FS 1983:3
Switzerland	SEV 3669, TM-Ph-046A, TM-Ph-048
U.K.	BS 7122
U.S.	21CFR1040, SSRL
U.S. (New York)	IND.CODE 50
U.S. (Alaska)	18 Chp 85,#7
U.S. (Arizona)	R12-1-1401
U.S. (Florida)	10D-89
U.S. (Georgia)	290-5-27
U.S. (Illinois)	P.A.81-1516
U.S. (Massachusetts)	105 CMR 21
Weiner	Safety Update
WHO	E.H.C.#2

Note: ANSI = American National Standards Institute. CENELEC = European Organization for Electrotechnical Standardization. IEC = International Electrotechnical Commission. WHO = World Health Organization.

of microwaves. Most of the reported effects are subjective, consisting of fatigue, headache, sleepiness, irritability, loss of appetite, and memory difficulties. Psychic changes that include unstable mood, hypochondriasis, and anxiety have been reported. The symptoms are reversible and pathological damage to neural structures is insignificant.

Most of the experimental work to the solve questions about the effects of electromagnetic waves upon organisms supports the belief that the chief effect on living tissue is to produce heating. Hence, exposure to microwave radiation should probably represent no hazard unless overheating is a possibility. Within carefully prescribed limits, the heating effect of radio waves may actually be beneficial.

Heating is a function of both the time of exposure and the strength of the microwave field, that is, the average power per unit area, usually expressed in milliwatts per square centimeter (mW/cm²). See Figure A.10. The heating may take place near the surface or deep within the body, the depth of penetration being related to frequency. Frequencies in the region 200–900 MHz penetrate deeply, whereas S-band (1.5–5.2 GHz) and X-band (5.2–11 GHz) produce heating at or near the surface.

Heating effects, depending on frequency, are (a) a general rise in body temperature or (b) something more localized in the body. As the surface of the human body is more generously supplied with sensory nerves than the interior, a feeling of warmth may give a warning in case of over-exposure to frequencies that produce surface heating. In the case of localized microwave heating deep within the body, it is still less likely that any warning sensations would be noted before damage was done.

Despite all this, it is important to establish limits and to delineate the areas in which a potential health hazard could exist.

A.9.1 Dosimetric Quantities

The greater part of research about microwave radiation hazards is based on both empiric measures of small animals (rabbits, dogs, mice, etc.) and models that simulate behaviors of human tissue. Measures were guided to find limits of security in the functions of power density, frequency, exposure time, and weight. At first results obtained from animals were accepted, although later clear differences between animals and humans were being observed.

The major point of contention appears to be the now widespread and accepted use of the specific absorption rate (SAR) in exposed tissue, rather than the previously used power density, as the important dosimetric quantity. This use of the SAR appears to be based on the fact that it is a measure of absorbed energy in the tissue, which is interpreted by the critics as being exactly equal to the heat generated in the exposed tissue:

$$E = \sqrt{\frac{10^3 * \rho * SAR}{\sigma}} \qquad (A.13)$$

where E is the electric field, σ is electrical conductivity, and ρ is the density of tissue.

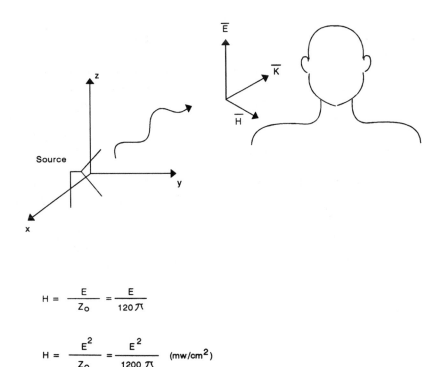

$$H = \frac{E}{Z_0} = \frac{E}{120\,\pi}$$

$$H = \frac{E^2}{Z_0} = \frac{E^2}{1200\,\pi} \quad (mw/cm^2)$$

Figure A.10 Quantities and units to specify the electromagnetic field exposure limits.

Since the SAR has been the easiest quantity to measure and is an index of the internal electric field strength, its use has been widely accepted. The measured electric field strength and known electrical conductivity of the tissue may be used to calculate the SAR by (A.13).

A.9.2 SAR in Application to Safety: Standards

The 1966 and 1974 American National Standards Institute (ANSI) guides were based on the best experimental evidence available at the time they were promulgated, as judged by a cross-section of scientists, many of whom are recognized as international authorities on the subject. The guides do not take into account the variation of energy coupling with frequency effects, and do not take into account the effects due to low-level exposure reported so frequently in Soviet and East European literature. So, ANSI C95 SC was created for the promulgation of a new Radio Frequency Protection Guide (RFPG) in 1980.

Today, methods to obtain safety limits are based on maximum allowable SAR for human exposure. Through a data selection of SAR it was possible to find an upper limit for all human sizes (Fig. A.11) [14].

Based on the relationship between incident power density and the total body average SAR, the engineers formulated frequency-dependent criteria for the standard in terms of a fixed maximum-allowed average SAR. The average SAR allowed for human exposure should be at least an order of magnitude below the threshold level of what they considered the threshold for hazardous effects. This resulted in the exposure criteria based on limiting the SAR to 0.4 W/kg, as shown in Figure A.12. and Table A.3.

It is interesting to compare the newly proposed ANSI whole-body exposure guide with the Soviet Occupational Standard and the Soviet General Population Standard, as illustrated in Figure A.13. The average power density for the Soviet magnetic field occupational standards are about three orders of magnitude greater than that for ANSI. This is exactly what one would expect if dosimetry quantities from magnetic field experiments on small animals are scaled for equivalence to human.

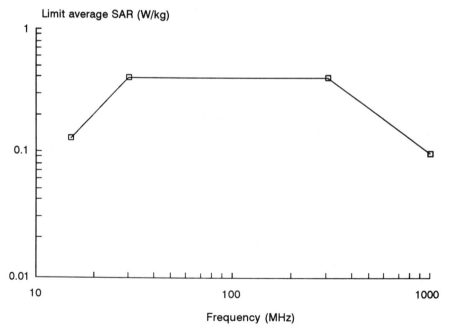

Figure A.11 Upper limit average SAR versus frequency for human exposure to electromagnetic plane-wave radiation.

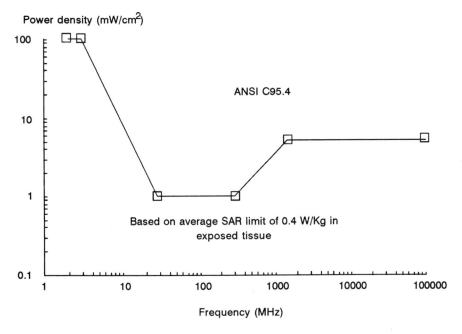

Figure A.12 Incident power density limit versus frequency.

Table A.3
ANSI Recommendation, Based on Frequency Dependence and SAR Limited to 0.4 W/Kg

Frequency (MHz)	Power Density (mW/cm²)	E² (V²/m²)	H² (A²/m²)
0.3–3	100	400,000	2.5
3–30	900/f²	4,000 (900/f²)	0.025 (900/f²)
30–300	1.0	4,000	0.025
300–1,500	f/300	4,000 (f/300)	0.025 (f/300)
1,500–100,000	5	20,000	0.125

A.9.3 Power Density and Minimum Safe Distance

To compute the value of the power density in the near field, assuming a circular dish antenna, use

$$W = \frac{16P}{\pi D^2} = 4\frac{P}{A} \tag{A.14}$$

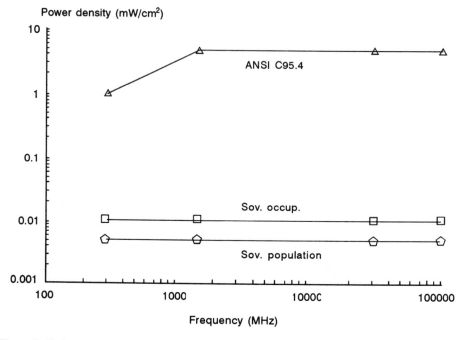

Figure A.13 Compared power density limit versus frequency.

where P = average power output (not peak power), D = diameter of antenna, and A = area of antenna. If this computation reveals a power density which is less than the limit, then there is no need to proceed with further calculations. If not, then one assumes that the limit may exist any place in the near-field region, and attention is directed to the far-field region. In the far-field region, the free-space power density on the beam axis may be computed from

$$W = \frac{GP}{4\pi r^2} = \frac{AP}{\lambda^2 r^2} \qquad (A.15)$$

where λ = wavelength. The distance from the antenna to the intersection of the near field (A.13) with the far field (A.14) is given by

$$r_1 = \frac{\pi D^2}{8\lambda} = \frac{A}{2\lambda} \qquad (A.16)$$

These formulas do not include the effect of ground reflection, which could cause a value of power density four times the free-space value. In this case (A.15) yields:

$$W = 4\,\frac{AP}{\lambda^2 r^2} \tag{A.17}$$

The formulas do neglect the loss of power due to "spill over" at the antenna and the effectiveness of the antenna area and, hence, will in general give conservative estimates. Moreover, transmitter powers are often rated as power available at the generator so that any transmission line losses between the generator and the antenna would make the estimates even more conservative.

For example, the effective area of a lossless loop or small dipole (doublet) is given by:

$$A_{eff} = \frac{3}{8}\,\frac{\lambda^2}{\pi} \tag{A.18}$$

With the equations above and Table A.3, it is possible to obtain the minimum safe distance where the average power density is equal to the ANSI standard in the function of average power transmitted at different frequencies (Fig. A.14).

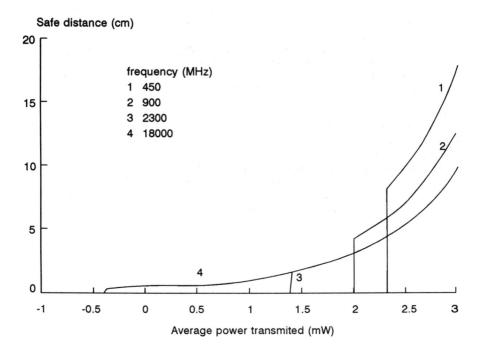

Figure A.14 Minimum safe distance versus average power transmitted for a small dipole.

REFERENCES

[1] Winburn, D. C. *Practical Laser Safety*, New York: Marcel Decker Inc., 1990.

[2] Sliney, D. H., and M. Wolbarsht. *Safety With Lasers and Other Optical Sources*, New York: Plenum Press, 1981.

[3] Ham, W. Y., H. A. Mueller, J. J. Ruffolo, R. K. Guerry, and A. M. Clarke. "Ocular Effect of GaAs Lasers and near Infrared Radiation," *Applied Optics*, Vol. 23, No. 13, July 1984.

[4] American National Standards Institute. *Safe Use of Lasers*, ANSI Z-136.1, Z-131.2 and Z-131.3.

[5] Landsberg, G. S. *Optic*, ed. MIR, Moscow, 1984.

[6] Tarasov, L., and A. Tarasova. *Light Refraction Lessons*, ed. MIR, Moscow, 1985.

[7] ACGIH-TLV. *American Conference of Governmental Industrial Hygienists—Threshold Limit Values*, 1979.

[8] ETI. *Laser Safety*, Engineering Technology Institute, Waco, 1983.

[9] Sliney, D. H. "Laser Safety Optical Radiation Hazards," Chap. 6 in *Handbook of Laser Science and Technology, Vol. 1*, edited by M. J. Weber, Boca Raton: CRC Press Inc., 1982.

[10] STC. *Product Safety Information Sheet*, 1985.

[11] Weiner, R. "Status of Laser Safety Requirements," *Laser & Optronics*, December 1991.

[12] Petersen, R. C., and D. H. Sliney. "Toward the Development of Laser Safety Standards for Fiber-Optic Communication Systems," *Applied Optics*, Vol. 25, No. 7, April 1986.

[13] Betancor, M. J., and V. M. Melian. *Final Report of ESPRIT WINS Project*. October 1991.

[14] Guy, A. W., "Non-ionizing Radiation: Dosimetry and Interaction," *Proc. of the Non-Ionizing Rad. Symp. ACGIH*, Washington, DC, November 26–28, 1979.

[15] Mumford, W. W. "Some Technical Aspects of Microwave Radiation Hazards," *IRE Proc.*, Vol. 49, February 1961, pp. 427–447.

[16] Tell, R.A., and F. Harlen. "A review of Selected Biological Effects and Dosimetric Data Useful for Development of Radiofrequency Safety Standards for Human Exposure," *J. of Microwave Power*, Vol. 14, No. 4, December 1979, pp. 405–424.

[17] American National Standards Institute. "Safety guide for the prevention of radio-frequency hazards in the use of electric blasting caps," *ANSI C95.4*, 1971.

[18] Michaelson, S. M. "Microwave/Radiofrequency Protection Standards: Concepts, Criteria and Applications," *Proc. if the 5th Intl. Congress of the Intl. Rad. Protection Assoc.*, Vol. 2, 1980, pp. 407–414.

[19] American National Standards Institute. "Safety of Electromagnetic Radiation with Respect to Personnel," *ANSI C95.1*, 1974.

[20] *Temporary Sanitary Rules for Working with Centimeter Waves*, Ministry of Health Protection of the USSR, USSR (1958).

[21] Marha, K. "Microwave Radiation Safety Standards in Eastern Europe," *IEEE Trans. on Microwave Theory and Techniques.*, Vol. MTT-19, February 1971, pp. 165–168.

[22] Janes Jr., D. E. "Radiofrequency Environments in the United States," *Proc. of the 15th IEEE Intl. Conf. on Communications*, Vol. 2, Boston, MA, June 10–14 1979, pp. 31.4.1–31.4.5.

Index

The Artech House Telecommunications Library

Vinton G. Cerf, Series Editor

Advances in Computer Communications and Networking, Wesley W. Chu, editor

Advances in Computer Systems Security, Rein Turn, editor

Analysis and Synthesis of Logic Systems, Daniel Mange

A Bibliography of Telecommunications and Socio-Economic Development, Heather E. Hudson

Codes for Error Control and Synchronization, Djimitri Wiggert

Communication Satellites in the Geostationary Orbit, Donald M. Jansky and Michel C. Jeruchim

Communications Directory, Manus Egan, editor

The Complete Guide to Buying a Telephone System, Paul Daubitz

The Corporate Cabling Guide, Mark W. McElroy

Corporate Networks: The Strategic Use of Telecommunications, Thomas Valovic

Current Advances in LANs, MANs, and ISDN, B. G. Kim, editor

Design and Prospects for the ISDN, G. Dicenet

Digital Cellular Radio, George Calhoun

Digital Hardware Testing: Transistor-Level Fault Modeling and Testing, Rochit Rajsuman, editor

Digital Signal Processing, Murat Kunt

Digital Switching Control Architectures, Giuseppe Fantauzzi

Digital Transmission Design and Jitter Analysis, Yoshitaka Takasaki

Distributed Processing Systems, Volume I, Wesley W. Chu, editor

Disaster Recovery Planning for Telecommunications, Leo A. Wrobel

Document Imaging Systems: Technology and Applications, Nathan J. Muller

E-Mail, Stephen A. Caswell

Enterprise Networking: Fractional T1 to SONET, Frame Relay to BISDN, Daniel Minoli

Expert Systems Applications in Integrated Network Management, E. C. Ericson, L. T. Ericson, and D. Minoli, editors

FAX: Digital Facsimile Technology and Applications, Second Edition, Dennis Bodson, Kenneth McConnell, and Richard Schaphorst

Fiber Network Service Survivability, Tsong-Ho Wu

Fiber Optics and CATV Business Strategy, Robert K. Yates et al.

A Guide to Fractional T1, J.E. Trulove

Handbook of Satellite Telecommunications and Broadcasting, L. Ya. Kantor, editor

Implementing X.400 and X.500: The PP and QUIPU Systems, Steve Kille

Inbound Call Centers: Design, Implementation, and Management, Robert A. Gable

Information Superhighways: The Economics of Advanced Public Communication Networks, Bruce Egan

Integrated Broadband Networks, Amit Bhargava

Integrated Services Digital Networks, Anthony M. Rutkowski

International Telecommunications Management, Bruce R. Elbert

International Telecommunication Standards Organizations, Andrew Macpherson

Internetworking LANs: Operation, Design, and Management, Robert Davidson and Nathan Muller

Introduction to Satellite Communication, Bruce R. Elbert

Introduction to T1/T3 Networking, Regis J. (Bud) Bates

Introduction to Telecommunication Electronics, A. Michael Noll

Introduction to Telephones and Telephone Systems, Second Edition, A. Michael Noll

Introduction to X.400, Cemil Betanov

The ITU in a Changing World, George A. Codding, Jr. and Anthony M. Rutkowski

Jitter in Digital Transmission Systems, Patrick R. Trischitta and Eve L. Varma

LAN/WAN Optimization Techniques, Harrell Van Norman

LANs to WANs: Network Management in the 1990s, Nathan J. Muller and Robert P. Davidson

The Law and Regulation of International Space Communication, Harold M. White, Jr. and Rita Lauria White

Long Distance Services: A Buyer's Guide, Daniel D. Briere

Mathematical Methods of Information Transmission, K. Arbenz and J. C. Martin

Measurement of Optical Fibers and Devices, G. Cancellieri and U. Ravaioli

Meteor Burst Communication, Jacob Z. Schanker

Minimum Risk Strategy for Acquiring Communications Equipment and Services, Nathan J. Muller

Mobile Information Systems, John Walker

Narrowband Land-Mobile Radio Networks, Jean-Paul Linnartz

Networking Strategies for Information Technology, Bruce Elbert

Numerical Analysis of Linear Networks and Systems, Hermann Kremer *et al.*

For further information on these and other Artech House titles, contact:

Artech House	Artech House
685 Canton Street	6 Buckingham Gate
Norwood, MA 01602	London SW1E6JP England
(617) 769-9750	+44(0)71 630-0166
Fax:(617) 762-9230	+44(0)71 630-0166
Telex: 951-659	Telex-951-659